U.S. Army Survival Series

The Complete Guide to
Shelter
Skills, Tactics, and Techniques

U.S. Army Survival Series

The Complete Guide to
Shelter
Skills, Tactics, and Techniques

Edited by
Jay McCullough

Skyhorse Publishing

NOV 0 1 2016

Copyright © 2007, 2016 by Skyhorse Publishing

All rights reserved. No part of this book may be reproduced in any manner without the express written consent of the publisher, except in the case of brief excerpts in critical reviews or articles. All inquiries should be addressed to Skyhorse Publishing, 307 West 36th Street, 11th Floor, New York, NY 10018. Skyhorse Publishing books may be purchased in bulk at special discounts for sales promotion, corporate gifts, fund-raising, or educational purposes. Special editions can also be created to specifications. For details, contact the Special Sales Department, Skyhorse Publishing, 307 West 36th Street, 11th Floor, New York, NY 10018 or info@skyhorsepublishing.com. Skyhorse® and Skyhorse Publishing® are registered trademarks of Skyhorse Publishing, Inc.®, a Delaware corporation.

Visit our website at www.skyhorsepublishing.com.

10 9 8 7 6 5 4 3 2 1

Library of Congress Cataloging-in-Publication Data is available on file.

Cover design by Tom Lau

Print ISBN: 978-1-5107-0742-9
Ebook ISBN: 978-1-5107-0747-4

Printed in the United States of America

CONTENTS

Foreword 6
Introduction 9
Chapter 1 Planning Positions 29
Chapter 2 Designing Positions 69
Chapter 3 Special Operations and Situations.....................81
Chapter 4 Position Design Details 101

Foreword

Survival is in many ways a matter of degree. An individual or group of people can survive the scarcity, or even complete removal, of many life-critical necessities, for a short time. But the more drastic the circumstances, over time, the chances for survival go down until the bitter end.

Shelter is critical to survival for a number of reasons. In acute circumstances, it can provide concealment, or effective protection from small-arms fire and explosives. Over the long run, it can augment survival probability by protecting against the loss of body heat, wind chill, bone-chilling precipitation and dew, the unrelenting rays of the sun, and by providing a comfortable place to sleep. Without effective shelter, a person is subject to radiant heat loss, convective wind chill, bullets, the discomfort and additional conductive heat loss from being wet, detection, sunburn, hunger from exposure, and to being harried by loss of life-sustaining sleep. Drip by drip, the chances for survival go down: survival, a matter of degrees.

In *The Complete Guide to U.S. Army Shelter Skills, Tactics, and Techniques,* we have a variety of U.S. Army-inspired shelter examples, from the most rudimentary and easily-constructed lean-to, to foxholes and Führerbunkers. In my estimation, the simplest examples are timelessand are the most useful to the outdoor sportsman or lost hiker. One would be well-advised to commit these to memory. Foxholes, trenches, and makeshift protection from bombs may also prove useful in unusual circumstances.

At the date of this writing, for a North American audience, it may seem to most readers that the construction of a Maginot Line-style defensive structure for a Mad Max–style hellscape would seem wholly unnecessary. Or, that if you need to build a bombproof bunker, a bug-out and rapid redeployment strategy may increase the odds of survival. But different readers have different necessities, and necessities for large populations can change with time. The especially vicious drug war in Mexico has driven many people to a bunker mentality; narcotraffickers, law enforcement, and innocent bystanders alike. It is not inconceivable that in an escalating bid for dominance—survivalevence—violence and social disorder of that type may eventually spill across the border. Reading the headlines, it is similarly conceivable that a financial disruption, or food or energy scarcity, or world events could drastically change our outlook. Laugh as we may at the nuclear bomb shelters of the 1950s, the armaments still remain. And recall that as recently as the 1940s, many Londoners relied on backyard bomb shelters during the blitz. The ordinary citizens of the 1930s could be forgiven for building backyard bomb shelters, when contemplating the folly of their betters leading up to World War II.

Despite this, some facts, as I see them, appear to be irrefutable for the foreseeable future. A thermonuclear world war does not appear to me to be survivable, or worth surviving. And preparing for an all-out assault, for any reason, against law enforcement or the military, is sure to be a losing proposition. The best and wisest shelter in such circumstances, would to be a dependable and tolerant neighbor, an advocate for justice, and to eschew war.

—Jay McCullough
North Haven, Connecticut
February 2016

Introduction

A shelter can protect you from the sun, insects, wind, rain, snow, hot or cold temperatures, and enemy observation. It can give you a feeling of well-being. It can help you maintain your will to survive. In some areas, your need for shelter may take precedence over your need for food and possibly even your need for water. For example, prolonged exposure to cold can cause excessive fatigue and weakness (exhaustion). An exhausted person may develop a "passive" outlook, thereby losing the will to survive. The most common error in making a shelter is to make it too large. A shelter must be large enough to protect you. It must also be small enough to contain your body heat, especially in cold climates.

SHELTER SITE SELECTION

When you are in a survival situation and realize that shelter is a high priority, start looking for shelter as soon as possible. As you do so, remember what you will need at the site. Two requisites are—

- It must contain material to make the type of shelter you need.
- It must be large enough and level enough for you to lie down comfortably.

When you consider these requisites, however, you cannot ignore your tactical situation or your safety. You must also consider whether the site—

- Provides concealment from enemy observation.
- Has camouflaged escape routes.
- Is suitable for signaling, if necessary.
- Provides protection against wild animals and rocks and dead trees that might fall.
- Is free from insects, reptiles, and poisonous plants.

You must also remember the problems that could arise in your environment.
For instance—

- Avoid flash flood areas in foothills.

- Avoid avalanche or rockslide areas in mountainous terrain.
- Avoid sites near bodies of water that are below the high water mark.

In some areas, the season of the year has a strong bearing on the site you select. Ideal sites for a shelter differ in winter and summer. During cold winter months you will want a site that will protect you from the cold and wind, but will have a source of fuel and water. During summer months in the same area you will want a source of water, but you will want the site to be almost insect free.

When considering shelter site selection, use the word BLISS as a guide.

B - Blend in with the surroundings.
L - Low silhouette.
I - Irregular shape.
S - Small.
S - Secluded location.

TYPES OF SHELTERS

When looking for a shelter site, keep in mind the type of shelter (protection) you need. However, you must also consider—

- How much time and effort you need to build the shelter.
- If the shelter will adequately protect you from the elements
- (sun, wind, rain, snow).
- If you have the tools to build it. If not, can you make improvised tools?
- If you have the type and amount of materials needed to build it.

To answer these questions, you need to know how to make various types of shelters and what materials you need to make them.

Poncho Lean-To

It takes only a short time and minimal equipment to build this lean-to (Figure I-1). You need a poncho, 2 to 3 meters of rope or parachute suspension line, three stakes about 30 centimeters long, and two trees or two poles 2 to 3 meters apart. Before selecting the trees you will use or the location of your poles, check the wind direction. Ensure that the back of your lean-to will be into the wind.

Introduction   11

Figure I-1: Poncho lean-to

To make the lean-to—

- Tie off the hood of the poncho. Pull the drawstring tight, roll the hood long ways, fold it into thirds, and tie it off with the drawstring.
- Cut the rope in half. On one long side of the poncho, tie half of the rope to the corner grommet. Tie the other half to the other corner grommet.
- Attach a drip stick (about a 10-centimeter stick) to each rope about 2.5 centimeters from the grommet. These drip sticks will keep rainwater from running down the ropes into the lean-to. Tying strings (about 10 centimeters long) to each grommet along the poncho's top edge will allow the water to run to and down the line without dripping into the shelter.
- Tie the ropes about waist high on the trees (uprights). Use a round turn and two half hitches with a quick-release knot.
- Spread the poncho and anchor it to the ground, putting sharpened sticks through the grommets and into the ground.

If you plan to use the lean-to for more than one night, or you expect rain, make a center support for the lean-to. Make this support with a line. Attach one end of the line to the poncho hood and the other end to an overhanging branch. Make sure there is no slack in the line.

Another method is to place a stick upright under the center of the lean-to. This method, however, will restrict your space and movements in the shelter.

For additional protection from wind and rain, place some brush, your rucksack, or other equipment at the sides of the lean-to.

To reduce heat loss to the ground, place some type of insulating material, such as leaves or pine needles, inside your lean-to.

Note: When at rest, you lose as much as 80 percent of your body heat to the ground.

To increase your security from enemy observation, lower the lean-to's silhouette by making two changes. First, secure the support lines to the trees at knee height (not at waist height) using two knee-high sticks in the two center grommets (sides of lean-to). Second, angle the poncho to the ground, securing it with sharpened sticks, as above.

Poncho Tent

This tent (Figure I-2) provides a low silhouette. It also protects you from the elements on two sides. It has, however, less usable space and observation area than a lean-to, decreasing your reaction time to enemy detection. To make this tent, you need a poncho, two 1.5- to 2.5-meter ropes, six sharpened sticks about 30 centimeters long, and two trees 2 to 3 meters apart.

Figure I-2. Poncho tent using overhanging branch

To make the tent—

- Tie off the poncho hood in the same way as the poncho lean-to.
- Tie a 1.5- to 2.5-meter rope to the center grommet on each side of the poncho.
- Tie the other ends of these ropes at about knee height to two trees 2 to 3 meters apart and stretch the poncho tight.
- Draw one side of the poncho tight and secure it to the ground pushing sharpened sticks through the grommets.
- Follow the same procedure on the other side.

If you need a center support, use the same methods as for the poncho lean-to. Another center support is an A-frame set outside but over the center of the tent (Figure I-3). Use two 90 to 120-centimeter-longsticks, one with a forked end, to form the A-frame. Tie the hood's drawstring to the A-frame to support the center of the tent.

Three-Pole Parachute Tepee

If you have a parachute and three poles and the tactical situation allows, make a parachute tepee. It is easy and takes very little time to make this tepee. It provides protection from the elements and can act as a signaling device by enhancing a small amount of light from a fire or candle. It is large enough to hold several people and their equipment and to allow sleeping, cooking, and storing firewood.

Figure I-3. Poncho tent with A-frame

You can make this tepee using parts of or a whole personnel main or reserve parachute canopy. If using a standard personnel parachute, you need three poles 3.5 to 4.5 meters long and about 5 centimeters in diameter.

To make this tepee (Figure I-4)—

- Lay the poles on the ground and lash them together at one end.
- Stand the framework up and spread the poles to form a tripod.
- For more support, place additional poles against the tripod. Five or six additional poles work best, but do not lash them to the tripod.
- Determine the wind direction and locate the entrance 90 degrees or more from the mean wind direction.
- Lay out the parachute on the "backside" of the tripod and locate the bridle loop (nylon web loop) at the top (apex) of the canopy.
- Place the bridle loop over the top of a free-standing pole. Then place the pole back up against the tripod so that the canopy's apex is at the same height as the lashing on the three poles.
- Wrap the canopy around one side of the tripod. The canopy should be of double thickness, as you are wrapping an entire parachute. You need only wrap half of the tripod, as the remainder of the canopy will encircle the tripod in the opposite direction.
- Construct the entrance by wrapping the folded edges of the canopy around two free-standing poles. You can then place the poles side by side to close the tepee's entrance.
- Place all extra canopy underneath the tepee poles and inside to create a floor for the shelter.
- Leave a 30- to 50-centimeter opening at the top for ventilation if you intend to have a fire inside the tepee.

One-Pole Parachute Tepee

You need a 14-gore section (normally) of canopy, stakes, a stout center pole, and inner core and needle to construct this tepee. You cut the suspension lines except for 40- to 45-centimeter lengths at the canopy's lower lateral band.

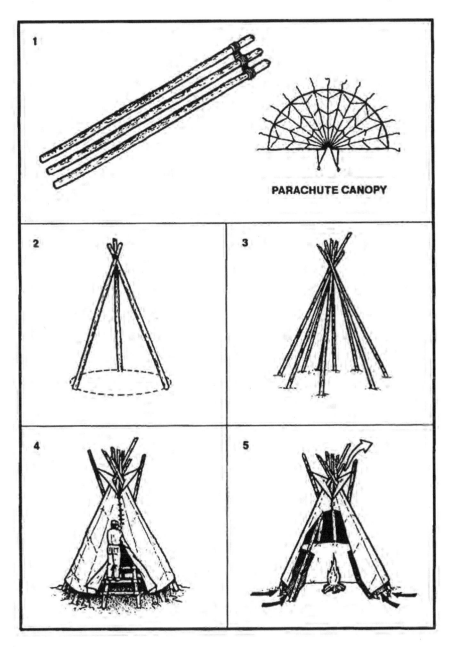

Figure I-4. Three-pole parachute tepee

To make this tepee (Figure I-5)—

- Select a shelter site and scribe a circle about 4 meters in diameter on the ground.
- Stake the parachute material to the ground using the lines remaining at the lower lateral band.
- After deciding where to place the shelter door, emplace a stake and tie the first line (from the lower lateral band) securely to it.
- Stretch the parachute material taut to the next line, emplace a stake on the scribed line, and tie the line to it.
- Continue the staking process until you have tied all the lines.
- Loosely attach the top of the parachute material to the center pole with a suspension line you previously cut and, through trial and error, determine the point at which the parachute material will be pulled tight once the center pole is upright.
- Then securely attach the material to the pole.
- Using a suspension line (or inner core), sew the end gores together Then securely attach the material to the pole.
- Using a suspension line (or inner core), sew the end gores together leaving 1 or 1.2 meters for a door.

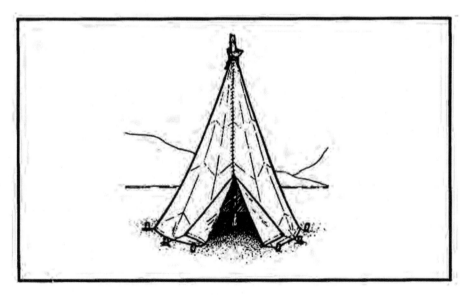

Figure I-5. One-pole parachute tepee

Introduction   17

Figure I-6. No-pole parachute tepee

No-Pole Parachute Tepee

You use the same materials, except for the center pole, as for the one-pole parachute tepee.

To make this tepee (Figure I-6)—
- Tie a line to the top of parachute material with a previously cut suspension line.
- Throw the line over a tree limb, and tie it to the tree trunk.
- Starting at the opposite side from the door, emplace a stake on the scribed 3.5- to 4.3-meter circle.
- Tie the first line on the lower lateral band.
- Continue emplacing the stakes and tying the lines to them.
- After staking down the material, unfasten the line tied to the tree trunk, tighten the tepee material by pulling on this line, and tie it securely to the tree trunk.

One-Man Shelter

A one-man shelter you can easily make using a parachute requires a tree and three poles. One pole should be about 4.5 meters long and the other two about 3 meters long.

To make this shelter (Figure I-7)—

- Secure the 4.5-meter pole to the tree at about waist height.

Figure I-7. One-man shelter

- Lay the two 3-meter poles on the ground on either side of and in the same direction as the 4.5-meter pole.
- Lay the folded canopy over the 4.5 meter pole so that about the same amount of material hangs on both sides.
- Tuck the excess material under the 3-meter poles, and spread it on the ground inside to serve as a floor.
- Stake down or put a spreader between the two 3-meter poles at the shelter's entrance so they will not slide inward.
- Use any excess material to cover the entrance.
- The parachute cloth makes this shelter wind resistant, and the shelter is small enough that it is easily warmed. A candle, used carefully, can keep the inside temperature comfortable. This shelter is unsatisfactory, however, when snow is falling as even a light snowfall will cave it in.

Parachute Hammock

You can make a hammock using 6 to 8 gores of parachute canopy and two trees about 4.5 meters apart (Figure I-8).

Field-Expedient Lean-To

If you are in a wooded area and have enough natural materials, you can make a field-expedient lean-to (Figure I-9) without the aid of tools or with only a knife. It takes longer to make this type of shelter than it does to make other types, but it will protect you from the elements.

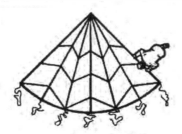

1 Lay out parachute and cut six gores of material.

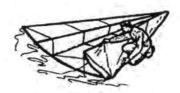

2 Starting from one side, make two folds each, one gore in width, yielding a base of three thicknesses of material.

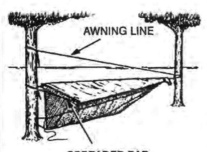

3 Suspend hammock between two trees with the skirt higher than the apex.* Place a spreader bar between the lines at the skirt and lace it to the skirt. Stretch an awning line between the two trees.

* An alternate and more stable configuration would be to tie each side of the skirt to a separate tree. However, this configuration of three trees could be difficult to find.

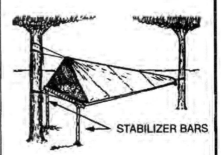

4 Drape the remaining three gores over the awning line and tuck the sixth gore into the shelter. Prop forked branches under the spreader bar to stabilize the shelter.

Figure I-8: Parachute hammock

You will need two trees (or upright poles) about 2 meters apart; one pole about 2 meters long and 2.5 centimeters in diameter; five to eight poles about 3 meters long and 2.5 centimeters in diameter for beams; cord or vines for securing the horizontal support to the trees; and other poles, saplings, or vines to crisscross the beams.

To make this lean-to—

- Tie the 2-meter pole to the two trees at waist to chest height. This is the horizontal support. If a standing tree is not available, construct a biped using Y-shaped sticks or two tripods.
- Place one end of the beams (3-meter poles) on one side of the horizontal support. As with all lean-to type shelters, be sure to place the lean-to's backside into the wind.
- Crisscross saplings or vines on the beams.
- Cover the framework with brush, leaves, pine needles, or grass, starting at the bottom and working your way up like shingling.
- Place straw, leaves, pine needles, or grass inside the shelter for Crisscross saplings or vines on the beams.
- Cover the framework with brush, leaves, pine needles, or grass, starting at the bottom and working your way up like shingling.
- Place straw, leaves, pine needles, or grass inside the shelter for bedding.

In cold weather, add to your lean-to's comfort by building a fire reflector wall (Figure I-9). Drive four 1.5-meter-long stakes into the ground to support the wall. Stack green logs on top of one another between the support stakes. Form two rows of stacked logs to create an inner space within the wall that you can fill with dirt. This action not only strengthens the wall but makes it more heat reflective. Bind the top of the support stakes so that the green logs and dirt will stay in place.

With just a little more effort you can have a drying rack. Cut a few 2-centimeter-diameter poles (length depends on the distance between the lean-to's horizontal support and the top of the fire reflector wall). Lay one end of the poles on the lean-to support and the other end on top of the reflector wall. Place and tie into place smaller sticks across these poles. You now have a place to dry clothes, meat, or fish.

Figure I-9. Field-expedient lean-to and fire reflector

Swamp Bed

In a marsh or swamp, or any area with standing water or continually wet ground, the swamp bed (Figure I-10) keeps you out of the water. When selecting such a site, consider the weather, wind, tides, and available materials.

To make a swamp bed—

- Look for four trees clustered in a rectangle, or cut four poles (bamboo is ideal) and drive them firmly into the ground so they form a rectangle. They should be far enough apart and strong enough to support your height and weight, to include equipment.
- Cut two poles that span the width of the rectangle. They, too, must be strong enough to support your weight.
- Secure these two poles to the trees (or poles). Be sure they are high enough above the ground or water to allow for tides and high water.
- Cut additional poles that span the rectangle's length. Lay them across the two side poles, and secure them.
- Cover the top of the bed frame with broad leaves or grass to form a soft sleeping surface.

Figure I-10: Swamp bed

- Build a fire pad by laying clay, silt, or mud on one corner of the swamp bed and allow it to dry.

Another shelter designed to get you above and out of the water or wet ground uses the same rectangular configuration as the swamp bed. You very simply lay sticks and branches lengthwise on the inside of the trees (or poles) until there is enough material to raise the sleeping surface above the water level.

Natural Shelters

Do not overlook natural formations that provide shelter. Examples are caves, rocky crevices, clumps of bushes, small depressions, large rocks on leeward sides of hills, large trees with low-hanging limbs, and fallen trees with thick branches. However, when selecting a natural formation—

- Stay away from low ground such as ravines, narrow valleys, or creek beds. Low areas collect the heavy cold air at night and are therefore colder than the surrounding high ground. Thick, brushy, low ground also harbors more insects.
- Check for poisonous snakes, ticks, mites, scorpions, and stinging ants.
- Look for loose rocks, dead limbs, coconuts, or other natural growth than could fall on your shelter.

Debris Hut

For warmth and ease of construction, this shelter is one of the best.
When shelter is essential to survival, build this shelter.
To make a debris hut (Figure I-11)—

- Build it by making a tripod with two short stakes and a long ridgepole or by placing one end of a long ridgepole on top of a sturdy base.
- Secure the ridgepole (pole running the length of the shelter) using the tripod method or by anchoring it to a tree at about waist height.
- Prop large sticks along both sides of the ridgepole to create a wedge-shaped ribbing effect. Ensure the ribbing is wide enough to accommodate your body and steep enough to shed moisture.
- Place finer sticks and brush crosswise on the ribbing. These form a latticework that will keep the insulating material

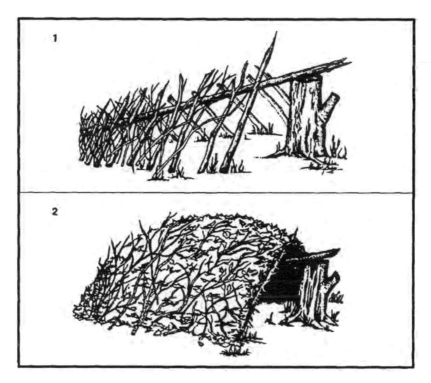

Figure I-11: Debris hut

(grass, pine needles, leaves) from falling through the ribbing into the sleeping area.
- Add light, dry, if possible, soft debris over the ribbing until the insulating material is at least 1 meter thick—the thicker the better.
- Place a 30-centimeter layer of insulating material inside the shelter.
- At the entrance, pile insulating material that you can drag to you once inside the shelter to close the entrance or build a door.
- As a final step in constructing this shelter, add shingling material or branches on top of the debris layer to prevent the insulating material from blowing away in a storm.

Tree-Pit Snow Shelter

If you are in a cold, snow-covered area where evergreen trees grow and you have a digging tool, you can make a tree-pit shelter (Figure I-12).

To make this shelter—

- Find a tree with bushy branches that provides overhead cover.

Figure I-12: Tree-pit snow shelter

- Dig out the snow around the tree trunk until you reach the depth and diameter you desire, or until you reach the ground.
- Pack the snow around the top and the inside of the hole to provide support.
- Find and cut other evergreen boughs. Place them over the top of the Pack the snow around the top and the inside of the hole to provide support.
- Find and cut other evergreen boughs. Place them over the top of the pit to give you additional overhead cover. Place evergreen boughs in the bottom of the pit for insulation.

See Chapter 3 for other arctic or cold weather shelters.

Beach Shade Shelter
This shelter protects you from the sun, wind, rain, and heat. It is easy to make using natural materials.
To make this shelter (Figure I-13)

- Find and collect driftwood or other natural material to use as support beams and as a digging tool.
- Select a site that is above the high water mark.
- Scrape or dig out a trench running north to south so that it receives the least amount of sunlight. Make the trench long and wide enough for you to lie down comfortably.

Figure I-13: Beach shade shelter

- Mound soil on three sides of the trench. The higher the mound, the more space inside the shelter.
- Lay support beams (driftwood or other natural material) that span the trench on top of the mound to form the framework for a roof.
- Enlarge the shelter's entrance by digging out more sand in front of it.
- Use natural materials such as grass or leaves to form a bed inside the shelter.

Desert Shelters
In an arid environment, consider the time, effort, and material needed to make a shelter. If you have material such as a poncho, canvas, or a parachute, use it along with such terrain features as rock outcropping, mounds of sand, or a depression between dunes or rocks to make your shelter.

Using rock outcroppings—

- Anchor one end of your poncho (canvas, parachute, or other material) on the edge of the outcrop using rocks or other weights.
- Extend and anchor the other end of the poncho so it provides the best possible shade.

In a sandy area—

- Build a mound of sand or use the side of a sand dune for one side of the shelter.
- Anchor one end of the material on top of the mound using sand or other weights.
- Extend and anchor the other end of the material so it provides the best possible shade.

Note: If you have enough material, fold it in half and form a 30-centimeter to 45-centimeter airspace between the two halves. This airspace will reduce the temperature under the shelter.

A belowground shelter (Figure I-14) can reduce the midday heat as much as 16 to 22 degrees C (30 to 40 degrees F). Building it, however, requires more time and effort than for other shelters. Since your

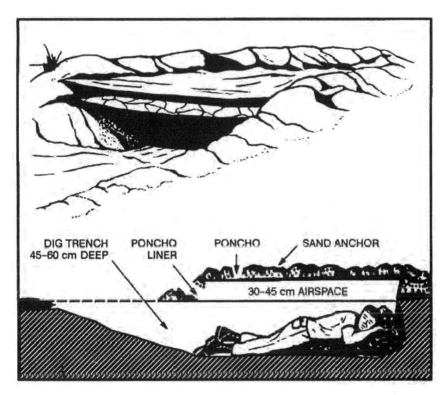

Figure I-14: Belowground desert shelter

physical effort will make you sweat more and increase dehydration, construct it before the heat of the day.

To make this shelter—

- Find a low spot or depression between dunes or rocks. If necessary, dig a trench 45 to 60 centimeters deep and long and wide enough for you to lie in comfortably.
- Pile the sand you take from the trench to form a mound around three sides.
- On the open end of the trench, dig out more sand so you can get in and out of your shelter easily.
- Cover the trench with your material.
- Secure the material in place using sand, rocks, or other weights.

If you have extra material, you can further decrease the midday temperature in the trench by securing the material 30 to 45 centimeters above the other cover. This layering of the material will reduce the inside temperature 11 to 22 degrees C (20 to 40 degrees F).

Another type of belowground shade shelter is of similar construction, except all sides are open to air currents and circulation. For maximum protection, you need a minimum of two layers of parachute material (Figure I-15). White is the best color to reflect heat; the innermost layer should be of darker material.

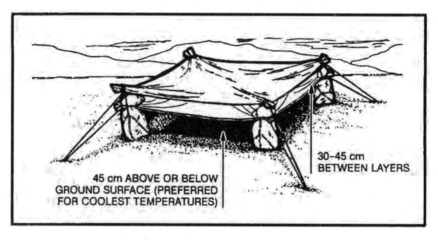

Figure I-15: Open desert shelter

CHAPTER 1

Planning Positions

This chapter highlights basic survivability knowledge required for planning fighting and protective positions. Included are descriptions f the various directly and indirectly fired weapons and their multiple penetration capabilities and effects on the positions. Both natural and man-made materials available to construct the positions are identified and ranked according to their protection potential. Positions are then categorized and briefly described. Construction methods, including the use of hand tools as well as explosives, and special overall construction considerations such as camouflage and concealment, are also presented.

WEAPONS EFFECTS

A fighting position is a place on the battlefield from which troops engage the enemy with direct and indirect fire weapons. The positions provide necessary protection for personnel, yet allow for fields of fire and maneuver. A protective position protects the personnel and/or material not directly involved with fighting the enemy from attack or environmental extremes. In order to develop plans for fighting and protective positions, five types of weapons, their effects, and their survivability considerations are presented. Air-delivered weapons such as ATGMs, laser-guided missiles, mines, and large bombs require similar survivability considerations.

DIRECT FIRE

Direct fire projectiles are primarily designed to strike a target with a velocity high enough to achieve penetration. The chemical energy projectile uses some form of chemical heat and blast to achieve penetration. It detonates either at impact or when maximum penetration is achieved. Chemical energy projectiles carrying impact-detonated or delayed detonation high-explosive charges are used mainly for direct fire from systems with high accuracy and consistently good target acquisition ability. Tanks, antitank weapons, and automatic cannons usually use these types of projectiles. The kinetic energy projectile uses high velocity and mass (momentum) to penetrate its target. Currently, the hypervelocity projectile causes the most concern in survivability position design. The materials used must dissipate the projectile's energy and thus prevent total penetration. Shielding against direct fire projectiles should initially stop or deform the projectiles in order to prevent or limit penetration. Direct fire projectiles are further divided into the categories of ball and tracer, armor piercing and armor piercing incendiary, and high explosive (HE) rounds.

Ball and Tracer
Ball and tracer rounds are normally of a relatively small caliber (5.56 to 14.5 millimeters (mm) and are fired from pistols, rifles, and machine guns. The round's projectile penetrates soft targets on impact at a high velocity. The penetration depends directly on the projectile's velocity, weight, and angle at which it hits.

Armor Piercing and Armor Piercing Incendiary
Armor piercing and armor piercing incendiary rounds are designed to penetrate armor plate and other types of homogeneous steel. Armor piercing projectiles have a special jacket encasing a hard core or penetrating rod which is designed to penetrate when fired with high accuracy at an angle very close to the perpendicular of the target. Incendiary projectiles are used principally to penetrate a target and ignite its contents. They are used effectively against fuel supplies and storage areas.

High Explosive
High explosive rounds include high explosive antitank (HEAT) rounds, recoilless rifle rounds, and antitank rockets. They are designed to detonate a shaped charge on impact. At detonation, an extremely high velocity molten jet is formed. This jet perforates large thicknesses of high-density material, continues along its path, and sets fuel and ammunition on fire. The HEAT rounds generally range in size from 60 to 120 mm.

Survivability Considerations
Direct fire survivability considerations include oblique impact, or impact of projectiles at other than a perpendicular angle to the structure, which increases the apparent thickness of the structure and decreases the possibility of penetration. The potential for ricochet off a structure increases as the angle of impact from the perpendicular increases. Designers of protective structures should select the proper

material and design exposed surfaces with the maximum angle from the perpendicular to the direction of fire. Also, a low structure silhouette design makes a structure harder to engage with direct fire.

INDIRECT FIRE
Indirect fire projectiles used against fighting and protective positions include mortar and artillery shells and rockets which cause blast and fragmentation damage to affected structures.

Blast
Blast, caused by the detonation of the explosive charge, creates a shock wave which knocks apart walls or roof structures. Contact bursts cause excavation cave-in from ground shock, or structure collapse. Overhead bursts can buckle or destroy the roof,

Blasts from high explosive shells or rockets can occur in three ways:

- Overhead burst (fragmentation from an artillery airburst shell).
- Contact burst (blast from an artillery shell exploding on impact).
- Delay fuze burst (blast from an artillery shell designed to detonate after penetration into a target).

The severity of the blast effects increases as the distance from the structure to the point of impact decreases. Delay fuze bursts are the greatest threat to covered structures. Repeated surface or delay fuze bursts further degrade fighting and protective positions by the cratering effect and soil discharge. Indirect fire blast effects also cause concussions. The shock from a high explosive round detonation causes headaches, nosebleeds, and spinal and brain concussions.

Fragmentation
Fragmentation occurs when the projectile disintegrates, producing amass of high-speed steel fragments which can perforate and become imbedded in fighting and protective positions. The pattern or distribution of fragments greatly affects the design of fighting and protective positions. Airburst of artillery shells provides the greatest unrestricted distribution of fragments. Fragments created by surface and delay bursts are restricted by obstructions on the ground.

Survivability Considerations
Indirect fire survivability from fragmentation requires shielding similar to that needed for direct fire penetration.

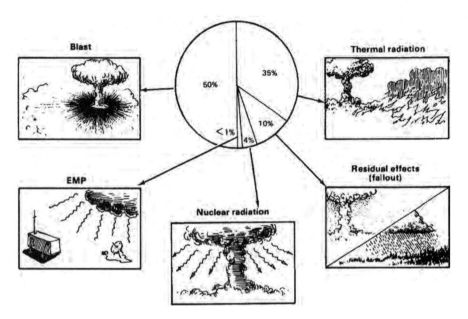

Figure 1-1: Energy distribution of tactical nuclear weapons

NUCLEAR

Nuclear weapons effects are classified as residual and initial. Residual effects (such as fallout) are primarily of long-term concern. However, they may seriously alter the operational plans in the immediate battle area. Figure 1-1 shows how the energy released by detonation of a tactical nuclear explosion is divided. Initial effects occur in the immediate area shortly after detonation and are the most tactically significant since they cause personnel casualties and material damage within the immediate time span of any operation. The principal initial casualty-producing effects are blast, thermal radiation (burning), and nuclear radiation. Other initial effects, such as electromagnetic pulse (EMP) and transient radiation effects on electronics (TREE), affect electrical and electronic equipment.

Blast
Blast from nuclear bursts overturns and crushes equipment, collapses lungs, ruptures eardrums, hurls debris and personnel, and collapses positions and structures.

Thermal Radiation
Thermal radiation sets fire to combustible materials, and causes flash blindness or burns in the eyes, as well as personnel casualties from skin burns.

Nuclear Radiation

Nuclear radiation damages cells throughout the body. This radiation damage may cause the headaches, nausea, vomiting, and diarrhea generally called "radiation sickness." The severity of radiation sickness depends on the extent of initial exposure. Table 1-1 shows the relationship between dose of nuclear radiation and distance from ground zero for a 1-kiloton weapon. Once the dose is known, initial radiation effects on personnel are determined from Table 1-2. Radiation in the body is cumulative.

Nuclear radiation is the dominant casualty-producing effect of low-yield tactical nuclear weapons. But other initial effects may produce significant damage and/or casualties depending on the weapon type, yield, burst conditions, and the degree of personnel and equipment protection. Figure 1-2 shows tactical radii of effects for nominal 1-kiloton and 10-kiloton weapons.

Electromagnetic Pulse

Electromagnetic pulse (EMP) damages electrical and electronic equipment. It occurs at distances from the burst where other nuclear weapons effects produce little or no damage, and it lasts for less than a second after the burst. The pulse also damages vulnerable electrical

Table 1-1: Relationship of radiation dose to distance from ground zero for a 1-KT weapon

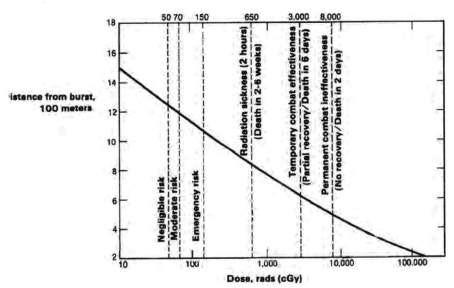

Table 1-2: Initial Radiation Effects on Personnel

Early Symptoms*

Dose rads (cGy)	Percent of Personnel	Time to Effect	Combat Effectiveness of Personnel	Fatalities
0 to 70	<5% of personnel require hospitalization		Full	None
150	5%	≤6 hours	Effectiveness reduced depending on task. Some hospitalization required.	None
650	100%	≤2 hours	Symptoms continue intermittently for next few days. Effectiveness reduced significantly for second to sixth day. Hospitalizaton required.	More than 50% in about 16 days
2,000 to 3,000	100%	≤5 minutes	Immediate, temporary incapacitation for 30 to 40 minutes, followed by recovery period during which efficiency is impaired. No operational capability.	100% in about 7 days
8,000	100%	≤5 minutes	Immediate, permanent incapacitation for personnel performing physically demanding tasks. No period of latent "recovery."	100% in 1 to 2 days
18,000	100%	Immediate	Permanent incapacitation for personnel performing even undemanding tasks. No operational capability.	100% within 24 hours

* Symptoms include vomiting, diarrhea, "dry heaving," nausea, lethargy, depression, and mental disorientation. At lower dose levels, incapacitation is a simple slow down in performance rate due to a loss of physical mobility and/or mental disorientation. At the high dose levels, shock and coma are sometimes the "early" symptoms.

and electronic equipment at ranges up to 5 kilometers for a 10-kiloton surface burst, and hundreds of kilometers for a similar high-altitude burst.

Survivability Considerations

Nuclear weapons survivability includes dispersion of protective positions within a suspected target area. Deep-covered positions will minimize the danger from blast and thermal radiation. Personnel should habitually wear complete uniforms with hands, face, and neck covered. Nuclear radiation is minimized by avoiding the radioactive

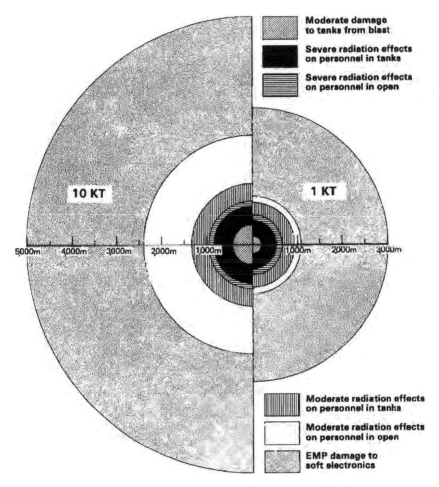

Figure 1-2: Tactical radii of effects of 1-KT and 10-KT fission weapons from low airburst

fallout area or remaining in deep-covered protective positions. Examples of expedient protective positions against initial nuclear effects are shown on Figure 1-3. Additionally, buttoned-up armor vehicles offer limited protection from nuclear radiation. Removal of antennae and placement of critical electrical equipment into protective positions will reduce the adverse effects of EMP and TREE.

CHEMICAL

Toxic chemical agents are primarily designed for use against personnel and to contaminate terrain and material. Agents do not destroy material and structures, bmake them unusable for periods

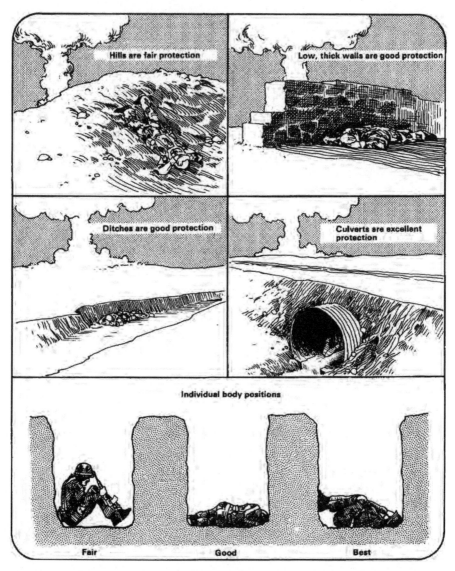

Figure 1-3: Examples of expedient protective positions against initial nuclear effects

of time because of chemical contaminant absorption. The duration of chemical agent effectiveness depends on—

- Weather conditions.
- Dispersion methods.
- Terrain conditions.

- Physical properties.
- Quantity used.
- Type used (nerve, blood, or blister).

Part II of this book provides chemical agent details and characteristics. Since the vapor of toxic chemical agents is heavier than air, it naturally tends to drift to the lowest corners or sections of a structure. Thus, low, unenclosed fighting and protective positions trap chemical vapors or agents. Because chemical agents saturate an area, access to positions without airlock entrance ways is limited during and after an attack, since every entering or exiting soldier brings contamination inside.

Survivability Considerations
Survivability of chemical effects includes overhead cover of any design that delays penetration of chemical vapors and biological aerosols, thereby providing additional masking time and protection against direct liquid contamination. Packing materials and covers are used to protect sensitive equipment. Proper use of protective clothing and equipment, along with simply avoiding the contaminated area, aids greatly in chemical survivability.

SPECIAL PURPOSE

Fuel-air munitions and flamethrowers are considered special-purpose weapons. Fuel-air munitions disperse fuel into the atmosphere forming a fuel-air mixture that is detonated. The fuel is usually contained in a metal canister and is dispersed by detonation of a central burster charge carried within the canister. Upon proper dispersion, the fuel-air mixture is detonated. Peak pressures created within the detonated cloud reach 300 pounds per square inch (psi). Fuel-air munitions create large area loading on a structure as compared to localized loadings caused by an equal weight high explosive charge. High temperatures ignite flammable materials. Flamethrowers and napalm produce intense heat and noxious gases which can neutralize accessible positions. The intense flame may also exhaust the oxygen content of inside air causing respiratory injuries to occupants shielded from the flaming fuel. Flame is effective in penetrating protective positions.

Survivability Considerations
Survivability of special purpose weapons effects includes covered positions with relatively small apertures and closable entrance areas which provide protection from napalm and flamethrowers.

Deep-supported tunnels and positions provide protection from other fuel-air munitions and explosives.

CONSTRUCTION MATERIALS

Before designing fighting and protective positions, it is important to know how the previously-described weapons affect and interact with various materials that are fired upon. The materials used in fighting and protective position construction act as either shielding for the protected equipment and personnel, structural components to hold the shielding in place, or both.

SHIELDING MATERIALS

Shielding provides protection against penetration of both projectiles and fragments, nuclear and thermal radiation, and the effects of fire and chemical agents. Various materials and amounts of materials provide varying degrees of shielding. Some of the more commonly used materials and the effects of both projectile and fragment penetration in these materials, as well as nuclear and thermal radiation suppression, are discussed in the following paragraphs. (Incendiary and chemical effects are generalized from the previous discussion of weapons effects.) The following three tables contain shielding requirements of various materials to protect against direct hits by direct fire projectiles (Table 1-3), direct fire high explosive (HE) shaped charges (Table 1-4), and indirect fire fragmentation and blast (Table 1-5). Table 1-6 lists nuclear protection factors associated with earth cover and sandbags.

Soil

Direct fire and indirect fire fragmentation penetration in soil or other similar granular material is based on three considerations: for materials of the same density, the finer the grain the greater the penetration; penetration decreases with increase in density; and penetration increases with increasing water content. Nuclear and thermal radiation protection of soil is governed by the following:

- The more earth cover, the better the shielding. Each layer of sandbags filled with sand or clay reduces transmitted radiation by 50 percent.
- Sand or compacted clay provides better radiation shielding than other soils which are less dense.
- Damp or wet earth or sand provides better protection than dry material.

Planning Positions 39

Table 1-3: Material thickness in inches, required to protect against direct hits by direct fire projectiles

Material	Small Caliber and Machine Gun (7.62-mm) Fire* at 100 yd	Antitank Rifle (76-mm) Fire at 100 yd	20-mm Antitank Fire at 200 yd	37-mm Antitank Fire at 400 yd	50-mm Antitank Fire at 400 yd	75-mm Direct Fire at 500 to 1,000 yd	Remarks
Solid walls**							
Brick masonry	18	24	30	60	-	-	None
Concrete, not reinforced***	12	18	24	42	48	54	Plain formed-concrete walls
Concrete, reinforced	6	12	18	36	42	48	Structurally reinforced with steel
Stone masonry	12	18	30	42	54	60	Values are guides only
Timber	36	60	-	-	-	-	Values are guides only
Wood	24	36	48	-	-	-	Values are guides only
Walls of loose material between boards**							
Brick rubble	12	24	30	60	72	-	None
Clay, dry	36	48	-	-	-	-	Add 100% to thickness if wet
Gravel/small crushed rock	12	24	30	60	72	-	None
Loam, dry	24	36	48	-	-	-	Add 50% to thickness if wet
Sand, dry	12	24	30	60	72	-	Add 100% to thickness if wet
Sandbags, filled with							
Brick rubble	20	30	30	60	70	-	None
Clay, dry	40	60	-	-	-	-	Add 100% to thickness if wet
Gravel/small crushed rock	20	30	30	60	70	-	None
Loam, dry	30	50	60	-	-	-	Add 50% to thickness if wet
Sand, dry	20	30	30	60	70	-	Add 100% to thickness if wet
Parapets of							
Clay	42	60	-	-	-	-	Add 100% to thickness if wet
Loam	36	48	60	-	-	-	Add 50% to thickness if wet
Sand	24	36	48	-	-	-	Add 100% to thickness if wet
Snow and Ice							
Frozen snow	80	80	-	-	-	-	None
Frozen soil	24	24	-	-	-	-	None
Icecrete (ice + aggregate)	18	18	-	-	-	-	None
Tamped snow	72	72	-	-	-	-	None
Unpacked snow	180	180	-	-	-	-	None

* One burst of five shots.
** Thicknesses to nearest ½ ft.
*** 3,000 psi concrete.

Note: Except where indicated, protective thicknesses are for a single shot only. Where weapons place five or six direct fire projectiles in the same area, the required protective thickness is approximately twice that indicated. Where no values are given, material is not recommended.

Table 1-4: Material thickness, in inches, required to protect against direct fire he shaped-charge

Material	73-mm RCLR	82-mm RCLR	85-mm RPG-7	107-mm RCLR	120-mm Sagger
Aluminum	36	24	30	36	36
Concrete	36	24	30	36	36
Granite	30	18	24	30	30
Rock	36	24	24	36	36
Snow, packed	156	156	156	--	--
Soil	100	66	78	96	96
Soil, frozen	50	33	39	48	48
Steel	24	14	18	24	24
Wood, dry	100	72	90	108	108
Wood, green	60	36	48	60	66

Note: Thicknesses assume perpendicular impact.

- Sandbags protected by a top layer of earth survive thermal radiation better than exposed bags. Exposed bags may burn, spill their contents, and become susceptible to the blast wave.

Steel

Steel is the most commonly used material for protection against direct and indirect fire fragmentation. Steel is also more likely to deform a projectile as it penetrates, and is much less likely to span than concrete. Steel plates, only 1/6 the thickness of concrete, afford equal protection against nondeforming projectiles of small and intermediate calibers. Because of its high density, steel is five times more effective in initial radiation suppression than an equal thickness of concrete. It is also effective against thermal radiation, although it transmits heat rapidly. Many field expedient types of steel are usable for shielding. Steel landing mats, culvert sections, and steel drums, for example, are effectively used in a structure as one of several composite materials. Expedient steel pieces are also used for individual protection against projectile and fragment penetration and nuclear radiation.

Concrete

When reinforcing steel is used in concrete, direct and indirect fire fragmentation protection is excellent. The reinforcing helps the concrete to remain intact even after excessive cracking caused by penetration, When a near-miss shell explodes, its fragments travel faster than its

Table 1-5: Material thickness, inches, required to protect against indirect fire fragmentation and blast exploding 50 feet away

Material	Mortars 82-mm	Mortars 120-mm	122-mm Rocket	HE Shells 122-mm	HE Shells 152-mm	Bombs 100-lb	Bombs 250-lb	Bombs 500-lb	Bombs 1,000-lb
Solid Walls									
Brick masonry	4	6	6	6	8	8	10	13	17
Concrete	4	5	5	5	6	8	10	15	18
Concrete, reinforced	3	4	4	4	5	7	9	12	15
Timber	8	12	12	12	14	15	18	24	30
Walls of loose material between boards									
Brick rubble	9	12	12	12	12	18	24	28	30
Earth*	12	12	12	12	16	24	30	-	-
Gravel, small stones	9	12	12	12	12	18	24	28	30
Sandbags, filled with									
Brick rubble	10	18	18	18	20	20	20	30	40
Clay*	10	18	18	18	20	30	40	40	50
Gravel, small stones, soil	10	18	18	18	20	20	20	30	40
Sand*	8	16	16	16	18	30	30	40	40
Loose parapets of									
Clay*	12	20	20	20	30	30	40	60	
Sand*	10	18	18	18	24	24	36	36	48
Snow									
Tamped	60	60	60	60	60	-	-	-	-
Unpacked	60	60	60	60	60	-	-	-	-

* Double values if material is saturated.
Note: Where no values are given, material is not recommended.

Table 1-6: Shielding values of earth cover and sandbags for a hypothetical 2,400-rads (cgy) free-in-air dose

Type of Protection	Radiation Protection Factor	Resulting Dose rads
Soldier in open	None	2,400
Earth Cover		
Soldier in 4-ft-deep open position	8	300
with 6 in of earth cover	12	200
with 12 in of earth cover	24	100
with 18 in of earth cover	48	50
with 24 in of earth cover	96	25
Sand- and Clay-Filled Sandbags		
Soldier in 4-ft-deep open position	8	300
with 1 layer of sandbags (4 in)	16	150
with 2 layers of sandbags (8 in)	32	75
with 3 layers of sandbags (12 in)	64	38

blast wave. If these fragments strike the exposed concrete surfaces of a protective position, they can weaken the concrete to such an extent that the blast wave destroys it. When possible, at least one layer of sandbags, placed on their short ends, or 15 inches of soil should cover all exposed concrete surfaces. An additional consequence of concrete penetration is spalling. If a projectile partially penetrates concrete shielding, particles and chunks of concrete often break or scab off the back of the shield at the time of impact. These particles can kill when broken loose. Concrete provides excellent protection against nuclear and thermal radiation.

Rock
Direct and indirect fire fragmentation penetration into rock depends on the rock's physical properties and the number of joints, fractures, and other irregularities contained in the rock. These irregularities weaken rock and can increase penetration. Several layers of irregularly-shaped rock can change the angle of penetration. Hard rock can cause a projectile or fragment to flatten out or break up and stop penetration. Nuclear and thermal radiation protection is limited because of undetectable voids and cracks in rocks. Generally, rock is not as effective against radiation as concrete, since the ability to provide protection depends on the rock's density.

Brick and Masonry
Direct and indirect fire fragmentation penetration into brick and masonry have the same protection limitations as rock. Nuclear and thermal radiation protection by brick and masonry is 1.5 times more effective than the protection afforded by soil. This characteristic is due to the higher compressive strength and hardness properties of brick and masonry. However, since density determines the degree of protection against initial radiation, unreinforced brick and masonry are not as good as concrete for penetration protection.

Snow and Ice
Although snow and ice are sometimes the only available materials in certain locations, they are used for shielding only. Weather could cause structures made of snow or ice to wear away or even collapse. Shielding composed of frozen materials provides protection from initial radiation, but melts if thermal radiation effects are strong enough.

Wood
Direct and indirect fire fragmentation protection using wood is limited because of its low density and relatively low compressive strengths.

Greater thicknesses of wood than of soil are needed for protection from penetration. Wood is generally used as structural support for a survivability position. The low density of wood provides poor protection from nuclear and thermal radiation. Also, with its low ignition point, wood is easily destroyed by fire from thermal radiation.

Other Materials
Expedient materials include steel pickets, landing mats, steel culverts, steel drums, and steel shipping consolidated express (CONEX) containers. Chapter 4 discusses fighting and protective positions constructed with some of these materials.

STRUCTURAL COMPONENTS
The structure of a fighting and protective position depends on the weapon or weapon effect it is designed to defeat. All fighting and protective positions have some configuration of floor, walls, and roof designed to protect material and/or occupants, The floor, walls, and roof support the shielding discussed earlier, or may in themselves make up that shielding, These components must also resist blast and ground shock effects from detonation of high explosive rounds which place greater stress on the structure than the weight of the components and the shielding. Designers must make structural components of the positions stronger, larger, and/or more numerous in order to defeat blast and ground shock, Following is a discussion of materials used to build floors, walls, and roofs of positions.

Floors
Fighting and protective position floors are made from almost any material, but require resistance to weathering, wear, and trafficability. Soil is most often used, yet is least resistant to water damage and rutting from foot and vehicle traffic. Wood pallets, or other field-available materials are often cut to fit floor areas. Drainage sumps, shown in Figure 1-4, or drains are also installed when possible.

Walls
Walls of fighting and protective positions are of two basic types—below ground (earth or revetted earth) and above-ground. Below-ground walls are made of the in-place soil remaining after excavation of the position. This soil may need revetment or support, depending on the soil properties and depth of cut. When used to support roof structures, earth walls must support the roof at points no less than one fourth the depth of cutout from the edges of excavation, as Figure 1-5.

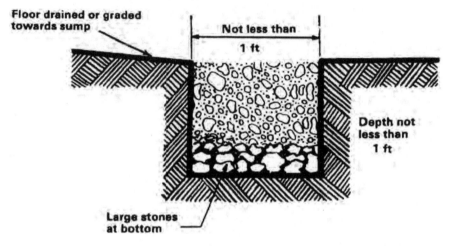

Figure 1-4: Drainage sump

Above-ground walls are normally constructed for shielding from direct fire and fragments. They are usually built of revetted earth, sandbags, concrete, or other materials. When constructed to a thickness adequate for shielding from direct fire and fragments, they are thick and stable enough for roof support.

Roofs

Roofs of fighting and protective positions are easily designed to support earth cover for shielding from fragments and small caliber direct fire. However, contact burst protection requires much stronger roof structures and, therefore, careful design. Roofs for support of earth cover shielding are constructed of almost any material that is usually used as beams or stringers and sheathing. Tables 1-7 and 1-8 present guidelines for wooden roof structures (for fragment shielding only). Table 1-9 converts dimensioned to round timber. Tables 1-10 and 1-11 pertain to steel pickets and landing mats for roof supports (for fragment shielding only).

When roof structures are designed to defeat contact bursts of high explosive projectiles, substantial additional roof protection is required. Table 1-13 gives basic design criteria for a roof to defeat contact bursts.

POSITION CATEGORIES

Seven categories of fighting and protective positions or components of positions that are used together or separately are—

Table 1-7: Maximum Span of Dimensioned Wood Roof Support for Earth Cover

Thickness of Earth Cover, ft	Span Length, ft					
	2½	3	3½	4	5	6
	Wood Thickness, in					
1½	1	1	2	2	2	2
2	1	2	2	2	2	3
2½	1	2	2	2	2	3
3	2	2	2	2	3	3
3½	2	2	2	2	3	3
4	2	2	2	2	3	4

Table 1-8: Maximum Span of Wood Stringer Roof Support for Earth Cover

Thickness of Earth Cover, ft	Span Length, ft					
	2½	3	3½	4	5	6
	Center-to-Center Spacing, in					
1½	40	30	22	16	10	18*
2	33	22	16	12	8/20*	14*
2½	27	18	12	10	16*	10*
3	22	14	10	8/20*	14*	8*
3½	18	12	8/24*	18*	12*	8*
4	16	10	8/20*	10*	10*	7*

Note: Stringers are 2 x 4s except those marked by an asterisk (*) which are 2 by 6s.

- Holes and simple excavations.
- Trenches.
- Tunnels.
- Earth parapets.
- Overhead cover and roof structures.
- Triggering screens.
- Shelters and bunkers.

Table 1-9: Converting Dimensioned Timber to Round Timber

4 x 4	5
6 x 6	7
6 x 8	8
8 x 8	10
8 x 10	11
10 x 10	12
10 x 12	13
12 x 12	14

*Sizes given are nominal and not rough cut timber.

Table 1-10: Maximum Span of Steel Picket Roof Supports for Sandbag Layers

Material	Triggering Requirements*
Plywood, dimensioned timber	2½-in thickness
Soil in sandbags with plywood or metal facing	2-in thickness (24-gage sheet metal)
Structured steel (corrugated metal)	¼-in thickness
Tree limbs	2-in diameter
Ammunition crates	1 layer (1-in-thick wood)
Snow	3 feet

* For detonating projectiles up to and including 120-mm mortar, rocket, and artillery shells.

Table 1-11: Maximum Span of Inverted Landing Mats (M8A1) for Roof Supports

Number of Sandbag Layers	Span Length, ft
2	10
5	6½
10	5
15	4
10	3½

HOLES AND SIMPLE EXCAVATIONS
Excavations, when feasible, provide good protection from direct fire and some indirect fire weapons effects. Open excavations have the advantages of—

- Providing good protection from direct fire when the occupant would otherwise be exposed.
- Permitting 360-degree observation and fire.
- Providing good protection from nuclear weapons effects.

Open excavations have the disadvantages of—

- Providing limited protection from direct fire while the occupant is firing a weapon, since frontal and side protection is negligible.
- Providing relatively no protection from fragments from overhead bursts of artillery shells. The larger the open excavation, the less the protection from artillery.
- Providing limited protection from chemical effects. In some cases, chemicals concentrate in low holes and excavations.

TRENCHES
Trenches provide essentially the same protection from conventional, nuclear, and chemical effects as the other excavations described, and are used almost exclusively in defensive areas. They are employed as protective positions and used to connect individual holes, weapons positions, and shelters. They provide protection and concealment for personnel moving between fighting positions or in and out of the area. They are usually open excavations, but sections are sometimes covered to provide additional protection. Trenches are difficult to camouflage and are easily detected from the air.

Trenches, like other positions, are developed progressively. As a general rule, they are excavated deeper than fighting positions to allow movement without exposure to enemy fire. It is usually necessary to provide revetment and drainage for them.

TUNNELS
Tunnels are not frequently constructed in the defense of an area due to the time, effort, and technicalities involved. However, they are usually used to good advantage when the length of time an area is defended justifies the effort, and the ground lends

itself to this purpose. The decision to build tunnels also depends greatly on the nature of the soil, which is usually determined by borings or similar means. Tunneling in hard rock is slow and generally impractical. Tunnels in clay or other soft soils are also impractical since builders must line them throughout to prevent collapse. Therefore, construction of tunneled defenses is usually limited to hilly terrain, steep hillsides, and favorable soils including hard chalk, soft sandstone, and other types of hard soil or soft rock.

In the tunnel system shown in Figure 1-5, the soil was generally very hard and only the entrances were timbered. The speed of excavation using hand tools varied according to the soil, and seldom exceeded 25 feet per day. In patches of hard rock, as little as 3 feet were excavated per day. Use of power tools did not significantly increase the speed of excavation. Engineer units, assisted by infantry personnel, performed the work. Tunnels of the type shown are excavated up to 30 feet below ground level. They are usually horizontal or nearly so. Entrances are strengthened against collapse under shell fire and ground shock from nuclear weapons. The first 16½ feet from

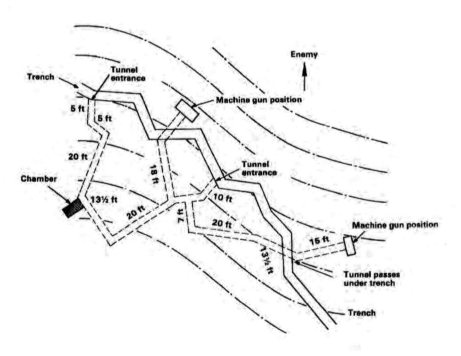

Figure 1-5: Earth wall roof support points

each entrance should have frames using 4 by 4s or larger timber supports.

Unlimbered tunnels are generally 3½ feet wide and 5 to 6½ feet high. Once beyond the portal or entrance, tunnels of up to this size are unlimbered if they are deep enough and the soil will stand open. Larger tunnels must have shoring. Chambers constructed in rock or extremely hard soil do not need timber supports. If timber is not used, the chamber is not wider than 6½ feet; if timbers are used, the width can increase to 10 feet. The chamber is generally the same height as the tunnel, and up to 13 feet long.

Grenade traps are constructed at the bottom of straight lengths where they slope. This is done by cutting a recess about 3½ feet deep in the wall facing the inclining floor of the tunnel.

Much of the spoil from the excavated area requires disposal and concealment. The volume of spoil is usually estimated as one third greater than the volume of the tunnel. Tunnel entrances need concealment from enemy observation. Also, it is sometimes necessary during construction to transport spoil by hand through a trench. In cold regions, air warmer than outside air may rise from a tunnel entrance thus revealing the position.

The danger that tunnel entrances may become blocked and trap the occupants always exists. Picks and shovels are placed in each tunnel so that trapped personnel can dig their way out, Furthermore, at least two entrances are necessary for ventilation. Whenever possible, one or more emergency exits are provided, These are usually small tunnels with entrances normally closed or concealed. A tunnel is constructed from inside the system to within a few feet of the surface so that an easy breakthrough is possible.

EARTH PARAPETS

Excavations and trenches are usually modified to include front, rear, and side earth parapets. Parapets are constructed using spoil from the excavation or other materials carried to the site. Frontal, side, and rear parapets greatly increase the protection of occupants firing their weapons (see Figure 1-6). Thicknesses required for parapets vary according to the material's ability to deny round penetration.

Parapets are generally positioned as shown below to allow full frontal protection, thus relying on mutual support of other firing positions. Parapets are also used as a single means of protection, even in the absence of excavations.

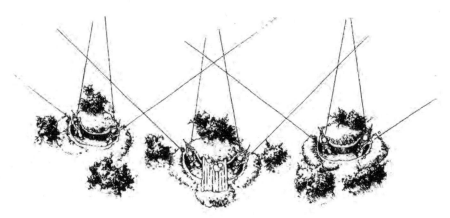

Figure 1-6: Parapets used for frontal protection relying on mutual support

OVERHEAD COVER AND ROOF STRUCTURES

Fighting and protective positions are given overhead cover primarily to defeat indirect fire projectiles landing on or exploding above them. Defeat of an indirect fire attack on a position, then, requires that the three types of burst conditions are considered. (Note: Always place a waterproof layer over any soil cover to prevent it from gaining moisture or weathering.)

Overhead Burst (Fragments)

Protection against fragments from airburst artillery is provided by a thickness of shielding required to defeat a certain size shell fragment, supported by a roof structure adequate for the dead load of the shielding. This type of roof structure is designed using the thicknesses to defeat fragment penetration given in Table 1-5. As a general guide, fragment penetration protection always requires at least 1 1/2 feet of soil cover. For example, to defeat fragments from a 120-mm mortar when available cover material is sandbags filled with soil, the cover depth required is 1 1/2 feet. Then, Table 1-8 shows that support of the 1 1/2 feet of cover (using 2 by 4 roof stringers over a 4-foot span) requires 16-inch center-to-center spacing of the 2 by 4s. This example is shown in Figure 1-7.

Contact Burst

Protection from contact burst of indirect fire HE shells requires much more cover and roof structure support than does protection from fragmentation. The type of roof structure necessary is given in Table 1-13. For example, if a position must defeat the contact burst of

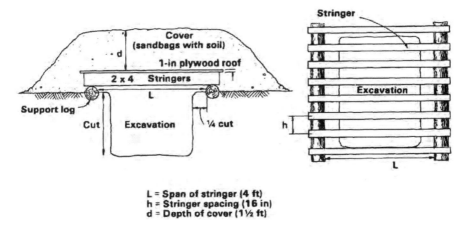

L = Span of stringer (4 ft)
h = Stringer spacing (16 in)
d = Depth of cover (1½ ft)

Figure 1-7: Position with overhead cover protection against fragments from a 120-mm mortar

an 82-mm mortar, Table 1-13 provides multiple design options. If 4 by 4 stringers are positioned on 9-inch center-to-center spacings over a span of 8 feet, then 2 feet of soil (loose, gravelly sand) is required to defeat the burst.

Delay Fuze Burst
Delay fuze shells are designed to detonate after penetration. Protection provided by overhead cover is dependent on the amount of cover remaining between the structure and the shell at the time of detonation. To defeat penetration of the shell, and thus cause it to detonate with a sufficient cover between it and the structure, materials are added on top of the overhead cover.

If this type of cover is used along with contact. burst protection, the additional materials (such as rock or concrete) are added in with the soil unit weight when designing the contact burst cover structure.

TRIGGERING SCREENS
Triggering screens are separately built or added on to existing structures used to activate the fuze of an incoming shell at a "standoff" distance from the structure. The screen initiates detonation at a distance where only fragments reach the structure. A variety of materials are usually used to detonate both super-quick fuzed shells and delay fuze shells up to and including 130 mm. Super-quick shell detonation requires only enough material to activate the fuze. Delay shells require more material to both limit penetration and activate the fuze. Typical standoff framing is shown below.

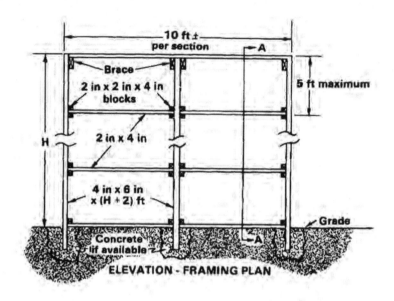

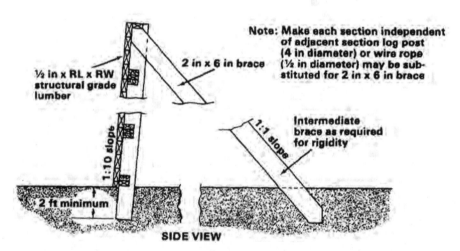

Figure 1-8: Typical standoff framing with dimensioned wood triggering screen

Defeating Super-Quick Fuzes

Incoming shells with super-quick fuzes are defeated at a standoff distance with several types of triggering screen materials. Table 1-12 lists thicknesses of facing material required for detonating incoming shells when impacting with the triggering screen. These triggering screens detonate the incoming shell but do not defeat fragments

Table 1-12: Triggering Screen Facing Material Requirements

Shells	Concrete*	Rock**	Rock Size (inches)
82-mm mortar	6	20	6½
120-mm mortar	20	36	9
122-mm rocket	50	40	10
122-mm artillery	68	40	10
130-mm artillery	80	42	10½

* 3,000 psi reinforced concrete.
** Rock must be relatively strong (compressive strength of about 20,000 psi) and in three layers for 82 mm; four layer for others.

Note: Due to the extreme thickness required for protection, materials such as earth, sand, and clay are not recommended.

from these shells. Protection from fragments is still necessary for a position. Table 1-13 lists required thicknesses for various materials to defeat fragments if the triggering screen is 10 feet from the structure.

Defeating Delay Fuzes

Delay fuzes are defeated by various thicknesses of protective material. Table 1-12 lists type and thickness of materials required to defeat penetration of delay fuze shells and cause their premature detonation, These materials are usually added to positions designed for contact burst protection. One method to defeat penetration and ensure premature shell detonation is to use layers of large stones. Figure 1-9 shows this added delay fuze protection on top of the contact burst protection. The rocks are placed in at least three layers on top of the required depth of cover for the expected shell size. The rock size is approximately twice the caliber of the expected shell. For example, the rock size required to defeat 82-mm mortar shell penetration is 2 x 82 mm = 164 mm (or 6 1/2 inches).

In some cases, chain link fences (shown below) also provide some standoff protection when visibility is necessary in front of the standoff and when positioned as shown in Figure 1-10. However, the fuze of some incoming shells may pass through the fence without initiating the firing mechanism.

SHELTERS AND BUNKERS

Protective shelters and fighting bunkers are usually constructed using a combination of the components of positions mentioned thus far. Protective shelters are primarily used as—

Table 1-13: Triggering Screen Material Thickness, in Inches, Required to Defeat Fragments at a 10-Foot Standoff

Nominal Stringer Size (inches)	Depth of Soil (d) (feet)	Center-to-Center Stringer Spacing (h) (inches), for Cited Span Length (L) (feet)				
		2	4	6	8	10
For Defeat of 82-mm Contact Burst						
2 x 4	2.0	3	4	4	4	3
	3.0	18	12	8	5	3
	4.0	18	14	7	4	3
2 x 6	2.0	4	7	8	8	6
	3.0	18	18	16	12	8
	4.0	18	18	18	11	7
4 x 4	2.0	7	10	10	9	7
	3.0	18	18	18	12	8
	4.0	18	18	18	10	7
4 x 8	1.5	4	5	7	8	8
	2.0	14	18	18	18	18
	3.0	18	18	18	18	18
For Defeat of 120- and 122-mm Contact Bursts						
4 x 8	2.0	-	-	-	-	-
	3.0	-	-	-	-	-
	4.0	3.5	4	5	5	6
	5.0	12	12	12	11	10
	6.0	18	18	18	16	12
6 x 6	2.0	-	-	-	-	-
	3.0	-	-	-	-	-
	4.0	-	-	5.5	6	6
	5.0	14	14	13	12	10
	6.0	18	18	18	16	12
6 x 8	2.0	-	-	-	-	-
	3.0	-	-	-	-	-
	4.0	5.5	6	8	9	10
	5.0	18	18	18	18	17
8 x 8	2.0	-	-	-	-	-
	3.0	-	-	-	-	-
	4.0	7.5	9	11	12	13
	5.0	18	18	18	18	18
For Defeat of 152-mm Contact Burst						
4 x 8	4.0	-	-	-	-	3.5
	5.0	6	6	7	7	7
	6.0	17	16	14	12	10
	7.0	18	18	18	15	11

(continued)

Table 1-13: *(Continued)*

6 x 6	4.0	-	-	-	-	-
	5.0	7	8	8	8	7
	6.0	18	18	15	12	10
	7.0	18	18	18	15	11
6 x 8	3.0	-	-	-	-	-
	4.0	-	-	-	-	6
	5.0	10	11	12	12	12
	6.0	18	18	18	18	17
8 x 8	3.0	-	-	-	-	-
	4.0	-	-	-	-	8
	5.0	14	15	16	17	16
	6.0	18	18	18	18	18

Table 1-14: Required Thickness, in Inches, of Protective Material to Resist Penetration of Different Shells (Delay Fuze)

Shells	Concrete*	Rock**	Rock Size (inches)
82-mm mortar	6	20	6½
120-mm mortar	20	36	9
122-mm rocket	50	40	10
122-mm artillery	68	40	10
130-mm artillery	80	42	10½

* 3,000 psi reinforced concrete.
** Rock must be relatively strong (compressive strength of about 20,000 psi) and in three layers for 82 mm; four layer for others.

Note: Due to the extreme thickness required for protection, materials such as earth, sand, and clay are not recommended.

- Command posts.
- Observation posts.
- Medical aid stations.
- Supply and ammunition shelters.
- Sleeping or resting shelters.

Protective shelters are usually constructed aboveground, using cavity wall revetments and earth-covered roof structures, or they are below ground using sections that are air transportable. Fighting

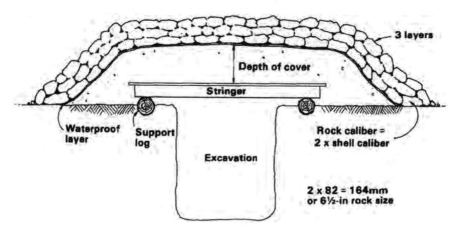

Figure 1-9: Stone layer added to typical overhead cover to defeat the delay fuze burst from an 82-mm mortar

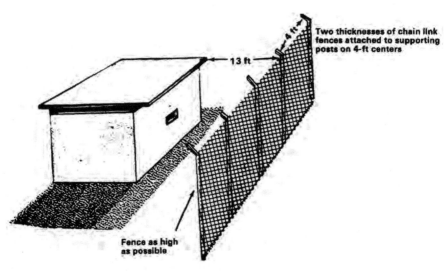

Figure 1-10: Chain link fence used for a standoff

bunkers are enlarged fighting positions designed for squad-size units or larger. They are built either aboveground or below ground and are usually made of concrete, However, some are prefabricated and transported forward to the battle area by trucks or air.

If shelters and bunkers are properly constructed with appropriate collective protection equipment, they can serve as protection against chemical and biological agents.

CONSTRUCTION METHODS

For individual and crew-served weapons fighting and protective position construction, hand tools are available. The individual soldier carries an entrenching tool and has access to picks, shovels, machetes, and hand carpentry tools for use in individual excavation and vertical construction work.

Earthmoving equipment and explosives are used for excavating protective positions for vehicles and supplies. Earthmoving equipment, including backhoes, bulldozers, and bucket loaders, are usually used for larger or more rapid excavation when the situation permits. Usually, these machines cannot dig out the exact shape desired or dig the amount of earth necessary. The excavation is usually then completed by hand, Descriptions and capabilities of US survivability equipment are given in appendix A.

Methods of construction include sandbagging, explosive excavation, and excavation revetments.

SANDBAGGING

Walls of fighting and protective positions are built of sandbags in much the same way bricks are used. Sandbags are also useful for retaining wall revetments as shown in Figure 1-11.

The sandbag is made of an acrylic fabric and is rot and weather resistant. Under all climatic conditions, the bag has a life of at least 2 years with no visible deterioration. (Some older-style cotton bags deteriorate much sooner.) The useful life of sandbags is prolonged by filling them with a mixture of dry earth and portland cement, normally in the ratio of 1 part of cement to 10 parts of dry earth. The cement sets as the bags take on moisture. A 1:6 ratio is used for sand-gravel mixtures. As an alternative, filled bags are dipped in a cement-water slurry, Each sandbag is then pounded with a flat object, such as a 2 by 4, to make the retaining wall more stable.

As a rule, sandbags are used for revetting walls or repairing trenches when the soil is very loose and requires a retaining wall. A sandbag revetment will not stand with a vertical face. The face must have a slope of 1:4, and lean against the earth it is to hold in place. The base for the revetment must stand on firm ground and dug at a slope of 4:1.

The following steps are used to construct a sandbag revetment wall such as the one shown in Figure 1-11.

- The bags are filled about three-fourths full with earth or a dry soil-cement mixture and the choke cords are tied.
- The bottom corners of the bags are tucked in after filling.

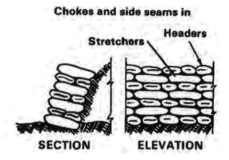

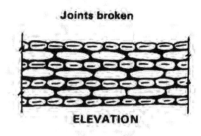

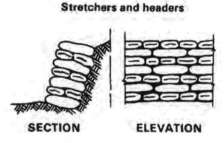

Figure 1-11: Retaining wall revetment

- The bottom row of the revetment is constructed by placing all bags as headers. The wall is built using alternate rows of stretchers and headers with the joints broken between courses. The top row of the revetment wall consists of headers.
- Sandbags are positioned so that the planes between the layers have the same pitch as the base—at right angles to the slope of the revetment.

- All bags are placed so that side seams on stretchers and choked ends on headers are turned toward the revetted face.
- As the revetment is built, it is backfilled to shape the revetted face to this slope.

Often, the requirement for filled sandbags far exceeds the capabilities of soldiers using only shovels. If the bags are filled from a stockpile, the job is performed easier and faster by using a lumber or steel funnel as shown in Figure 1-12.

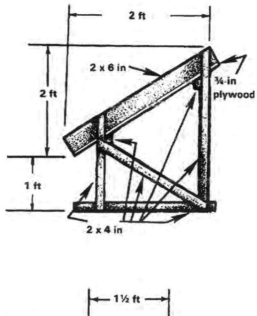

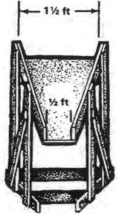

Figure 1-12: Expedient funnel for filling sandbags

EXCAVATION REVETMENTS

Excavations in soil may require revetment to prevent side walls from collapsing. Several methods of excavation revetments are usually used to prevent wall collapse.

Wall Sloping

The need for revetment is sometimes avoided or postponed by sloping the walls of the excavation. In most soils, a slope of 1:3 or 1:4 is sufficient. This method is used temporarily if the soil is loose and no revetting materials are available. The ratio of 1:3, for example, will determine the slope by moving 1 foot horizontally for each 3 feet vertically. When wall sloping is used, the walls are first dug vertically and then sloped.

Facing Revetments

Facing revetments serve mainly to protect revetted surfaces from the effects of weather and occupation. It is used when soils are stable enough to sustain their own weight. This revetment consists of the revetting or facing material and the supports which hold the revetting material in place. The facing material is usually much thinner than that used in a retaining wall. Facing revetments are preferable to wall sloping since less excavation is required. The top of the facing is set below ground level. The facing is constructed of brushwood hurdles, continuous brush, poles, corrugated metal, plywood, or burlap and chicken wire. The following paragraphs describe the method of constructing each type.

Brushwood Hurdle (Figure 1-13). A brushwood hurdle is a woven revetment unit usually 6 1/2 feet long and as high as the revetted wall. Pieces of brushwood about 1 inch in diameter are weaved on a framework of sharpened pickets driven into the ground at 20-inch intervals. When completed, the 6 1/2-foot lengths are carried to the position where the pickets are driven in place. The tops of the pickets are tied back to stakes or holdfasts and the ends of the hurdles are wired together.

Continuous Brush (Figure 1-14). A continuous brush revetment is constructed in place. Sharpened pickets 3 inches in diameter are driven into the bottom of the trench at 30-inch intervals and about 4 inches from the revetted earth face. The space behind the pickets is packed with small, straight brushwood laid horizontally. The tops of the pickets are anchored to stakes or holdfasts.

Pole (Figure 1-15). A pole revetment is similar to the continuous brush revetment except that a layer of small horizontal round poles,

Planning Positions 61

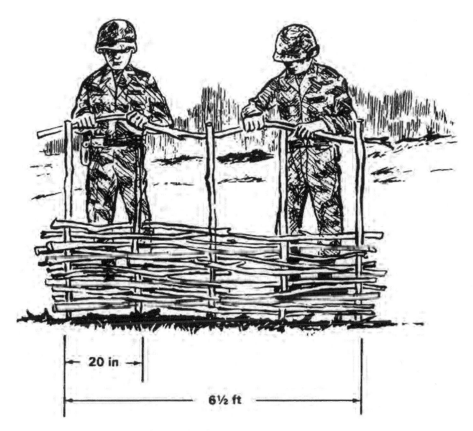

Figure 1-13: Brushwood hurdle

cut to the length of the revetted wall, is used instead of brushwood. If available, boards or planks are used instead of poles because of quick installation. Pickets are held in place by holdfasts or struts.

Corrugated Metal Sheets or Plywood (Figure 1-16). A revetment of corrugated metal sheets or plywood is usually installed rapidly and is strong and durable. It is well adapted to position construction because the edges and ends of sheets or planks are lapped, as required, to produce a revetment of a given height and length. All metal surfaces are smeared with mud to reduce possible reflection of thermal radiation and aid in camouflage. Burlap and chicken wire revetments are similar to revetments made from corrugated metal sheets or plywood. However, burlap and chicken wire does not have the strength or durability of plywood or sheet metal in supporting soil.

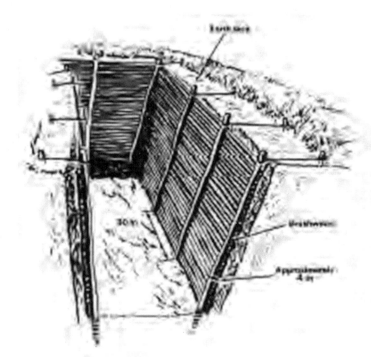

Figure 1-14: Continuous brush revetment

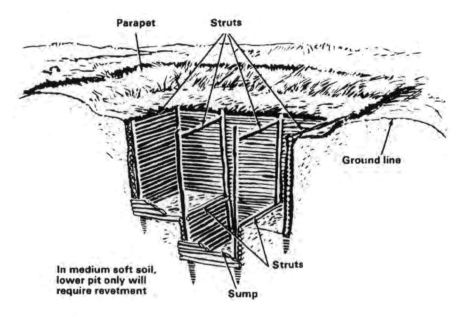

Figure 1-15: Pole revetment

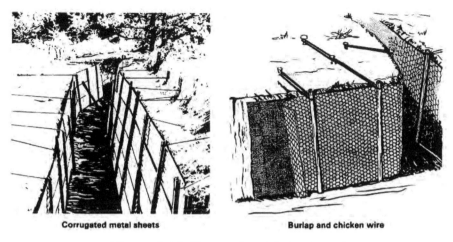

Corrugated metal sheets   Burlap and chicken wire

Figure 1-16: Types of metal revetment

Methods to Support Facing
The revetment facing is usually supported by timber frames (Figure 1-17) or pickets (Figure 1-18). Frames of dimensioned timber are constructed to fit the bottom and sides of the position and hold the facing material apart over the excavated width.

Pickets are driven into the ground on the position side of the facing material. The pickets are held tightly against the facing by bracing them apart across the width of the position. The size of pickets required and their spacing are determined by the soil and type of facing material used. Wooden pickets smaller than 3 inches in diameter are not used. The maximum spacing between pickets is about 6 1/2 feet. The standard pickets used to support barbed wire entanglements are excellent for use in revetting. Pickets are driven at least 1 1/2 feet into the floor of the position. Where the tops of the pickets are anchored, an anchor stake or holdfast is driven into the top of the bank and tied to the top of the picket. The distance between the anchor stake and the facing is at least equal to the height of the revetted face, with alternate anchors staggered and at least 2 feet farther back. Several strands of wire holding the pickets against the emplacement walls are placed straight and taut. A groove or channel is cut in the parapet to pass the wire through.

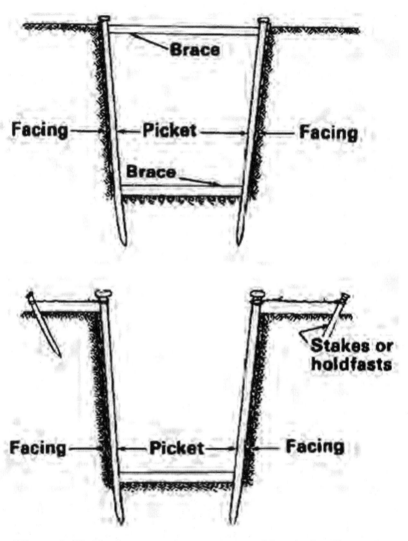

Figure 1-17: Facing revetment supported by timber frames

SPECIAL CONSTRUCTION CONSIDERATIONS

CAMOUFLAGE AND CONCEALMENT

The easiest and most efficient method of preventing the targeting and destruction of a position or shelter is use of proper camouflage and concealment techniques. Following are some general guidelines for position construction.

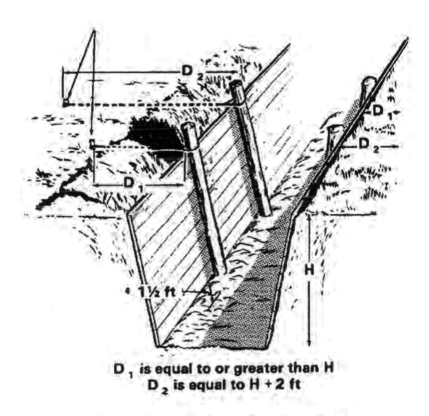

Figure 1-18: Facing revetment supported by pickets

Natural concealment and good camouflage materials are used. When construction of a positions begins, natural materials such as vegetation, rotting leaves, scrub brush, and snow are preserved for use as camouflage when construction is completed. If explosive excavation is used, the large area of earth spray created by detonation is camouflaged or removed by first placing tarpaulins or scrap canvas on the ground prior to charge detonation. Also, heavy equipment tracks and impressions are disguised upon completion of construction.

Fields of fire are not overcleared. In fighting position construction, clearing of fields of fire is an important activity for effective engagement of the enemy. Excessive clearing is prevented in order to reduce early enemy acquisition of the position. Procedures for clearing allow for only as much terrain modification as is needed for enemy acquisition and engagement.

Concealment from aircraft is provided. Consideration is usually given to observation from the air. Action is taken to camouflage position interiors or roofs with fresh natural materials, thus preventing contrast with the surroundings.

During construction, the position is evaluated from the enemy side. By far, the most effective means of evaluating concealment and camouflage is to check it from a suspected enemy avenue of approach.

DRAINAGE

Positions and shelters are designed to take advantage of the natural drainage pattern of the ground. They are constructed to provide for—

- Exclusion of surface runoff.
- Disposal of direct rainfall or seepage.
- Bypassing or rerouting natural drainage channels if they are intersected by the position.

In addition to using materials that are durable and resistant to weathering and rot, positions are protected from damage due to surface runoff and direct rainfall, and are repaired quickly when erosion begins. Proper position siting can lessen the problem of surface water runoff. Surface water is excluded by excavating intercepted ditches uphill from a position or shelter. Preventing water from flowing into the excavation is easier than removing it. Positions are located to direct the runoff water into natural drainage lines. Water within a position or shelter is carried to central points by constructing longitudinal slopes in the bottom of the excavation. A very gradual slope of 1 percent is desirable.

MAINTENANCE

If water is allowed to stand in the bottom of an excavation, the position is eventually undermined and becomes useless. Sumps and drains are kept clean of silt and refuse. Parapets around positions are kept clear and wide enough to prevent parapet soil from falling into the excavation, When wire and pickets are used to support revetment material, the pickets may become loose, especially after rain. Improvised braces are wedged across the excavation, at or near floor level, between two opposite pickets. Anchor wires are tightened by further twisting. Anchor pickets are driven in farther to hold tightened wires. Periodic inspections of sandbags are made.

REPAIRS

If the walls are crumbling in at the top of an excavation (ground level), soil is cut out where it is crumbling (or until firm soil is reached). Sandbags or sod blocks are used to build up the damaged area, If excavation walls are wearing away at the floor level, a plank is placed on its edge or the brushwood is shifted down. The plank is held against the excavation wall with short pickets driven into the floor. If planks are used on both sides of the excavation, a wedge is placed between the planks and earth is placed in the back of the planks. If an entire wall appears ready to collapse, the excavation is completely revetted. See Figure 1-19.

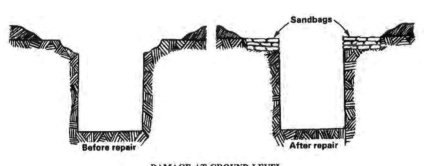

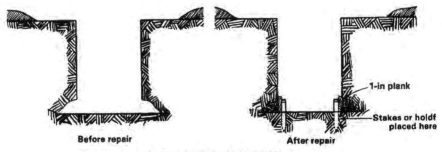

Figure 1-19: Excavation repair

SECURITY

In almost all instances, fighting and protective positions are prepared by teams of at least two personnel, During construction, adequate frontal and perimeter protection and observation are necessary. Additional units are sometimes required to secure an area during position construction. Unit personnel can also take turns with excavating and providing security.

CHAPTER 2

Designing Positions

This chapter contains basic requirements which must be built into the designs of fighting and protective positions. These requirements ensure soldiers are well-protected while performing their missions. The positions are all continuously improved as time, assets, and the situation permit. The following position categories are presented: hasty and deliberate fighting position for individual soldiers; trenches connecting the positions; positions for entire units; and special designs including shelters and bunkers. The positions in each category are briefly described and accompanied by a typical design illustration. Each category is summarized providing time and equipment estimates and protection factors for each position. Complete detailed construction drawings, and time and material estimates for a variety of positions are contained in Chapter 4.

A] BASIC DESIGN REQUIREMENTS

WEAPON EMPLOYMENT
While it is desirable for a fighting position to give maximum protection to personnel and equipment, primary consideration is always given to effective weapon use. In offensive combat operations, weapons are sited wherever natural or existing positions are available, or where weapon emplacement is made with minimal digging.

COVER
Positions are designed to defeat an anticipated threat. Protection against direct and indirect fire is of primary concern for position design. However, the effects of nuclear and chemical attack are taken into consideration if their use is suspected. Protection design for one type of enemy fire is not necessarily effective against another. The following three types of cover— frontal, overhead, and flank and rear— will have a direct bearing on designing and constructing positions.

Frontal
Frontal cover provides protection from small caliber direct fire. Natural frontal protection such as large trees, rocks, logs, and rubble is

best because enemy detection of fighting positions becomes difficult. However, if natural frontal protection is not adequate for proper protection, dirt excavated from the position (hole) is used. Frontal cover requires the position to have the correct length so that soldiers have adequate room; the correct dirt thickness (3 feet) to stop enemy small caliber fire; the correct height for overhead protection; and, for soldiers firing to the oblique, the correct frontal distance for elbow rests and sector stakes. Protection from larger direct fire weapons (for example, tank guns) is achieved by locating the position where the enemy cannot engage it, and concealing it so pinpoint location is not possible. Almost twice as many soldiers are killed or wounded by small caliber fire when their positions do not have frontal cover.

Overhead
Overhead cover provides protection from indirect fire fragmentation. When possible, overhead cover is always constructed to enhance protection against airburst artillery shells. Overhead cover is necessary because soldiers are at least ten times more protected from indirect fire if they are in a hole with overhead cover.

Flank and Rear
Flank and rear cover ensures complete protection for fighting positions, Flank and rear cover protects soldiers against the effects of indirect fire bursts to the flanks or rear of the position, and the effects of friendly weapons located in the rear (for example, packing from discarded sabot rounds fired from tanks). Ideally, this protection is provided by natural cover. In its absence, a parapet is constructed as time and circumstances permit.

SIMPLICITY AND ECONOMY
The position is usually uncomplicated and strong, requires as little digging as possible, and is constructed of immediately available materials.

INGENUITY
A high degree of imagination is essential to assure the best use of available materials. Many different materials existing on the battlefield and prefabricated materials found in industrial and urban areas can be used for position construction.

PROGRESSIVE DEVELOPMENT
Positions should allow for progressive development to insure flexibility, security, and protection in depth. Hasty positions are continuously improved into deliberate positions to provide maximum

protection from enemy fire. Trenches or tunnels connecting fighting positions give ultimate flexibility in fighting from a battle position or strongpoint. Grenade sumps are usually dug at the bottom of a position's front wall where water collects. The sump is about 3 feet long, 1/2 foot wide, and dug at a 30-degree angle. The slant of the floor channels excess water and grenades into the sump. In larger positions, separate drainage sumps or water drains are constructed to reduce the amount of water collecting at the bottom of the position.

CAMOUFLAGE AND CONCEALMENT

Camouflage and concealment activities are continual during position siting preparation. If the enemy cannot locate a fighting position, then the position offers friendly forces the advantage of firing first before being detected.

INDIVIDUAL FIGHTING POSITIONS

Table 2-1 summarizes the hasty and deliberate individual fighting positions and provides time estimates, equipment requirements, and protection factors.

HASTY POSITIONS

When time and materials are limited, troops in contact with the enemy use a hasty fighting position located behind whatever cover is available. It should provide frontal protection from direct fire while allowing fire to the front and oblique. For protection from indirect fire, a hasty fighting position is located in a depression or hole at least 1 1/2 feet deep. The following positions provide limited protection and are used when there is little or no natural cover. If the unit remains in the area, the hasty positions are further developed into deliberate positions which provide as much protection as possible.

DELIBERATE POSITIONS

Deliberate fighting positions are modified hasty positions prepared during periods of relaxed enemy pressure. If the situation permits, the unit leader verifies the sectors of observation before preparing each position. Continued improvements are made to strengthen the position during the period of occupation. Small holes are dug for automatic rifle biped legs so the rifle is as close to ground level as possible. Improvements include adding overhead cover, digging trenches to adjacent positions, and maintaining camouflage.

TRENCHES

Trenches are excavated to connect individual fighting positions and weapons positions in the progressive development of a defensive area. They provide protection and concealment for personnel moving between fighting positions or in and out of the area. Trenches are usually included in the overall layout plan for the defense of a position or strongpoint, Excavating trenches involves considerable time, effort, and materials, and is only justified when an area is occupied for a long time. Trenches are usually open excavations, but covered sections provide additional protection if the overhead cover does not interfere with the fire mission of the occupying personnel. Trenches are difficult to camouflage and are easily detected, especially from the air.

Table 2-1: Characteristics of individual fighting positions

Type of Position	Estimated Construction Time (man-hours)	Equipment Requirements	Direct Small Caliber Fire	Indirect Fire Blast and Fragmentation (Near-Miss)*	Indirect Fire Blast and Fragmentation (Direct Hit)	Nuclear Weapons**	Remarks
Hasty							
Crater	0.2	Hand tools	7.62mm	Better than in open - no overhead protection	None	Fair	
Skirmisher's trench	0.5	Hand tools	7.62mm	Better than in open - no overhead protection	None	Fair	
Prone position	1.0	Hand tools	7.62mm	Better than in open - no overhead protection	None	Fair	Provides all-around cover
Deliberate							
One-soldier position	3.0	Hand tools	12.7mm	Medium artillery no closer than 30 ft - no overhead protection	None	Fair	
One-soldier position with 1 1/2 ft. overhead cover	8.0	Hand tools	12.7mm	Medium artillery no closer than 30 ft	None	Good	Additional cover provides protection from direct hit small mortar blast
Two-soldier position	6.0	Hand tools	12.7mm	Medium artillery no closer than 30 ft - no overhead protection	None	Fair	
Two-Soldier position with 1 1/2 ft. overhead cover	11.0	Hand tools	12.7mm	Medium artillery no closer than 30 ft	None	Good	Additional cover provides protection from direct hit small mortar blast
LAW position	3.0	Hand tools	12.7mm	Medium artillery no closer than 30 ft - no overhead protection	None	Fair	

Note: Chemical protection is assumed because of individual protective masks and clothing.

* Shell sizes are: Small   Medium
   Mortar   82mm   120mm
   Artillery 105mm 152mm

** Nuclear protection ratings are rated poor, fair, good, very good, and excellent

Table 2-2: Shielding of m8a1 landing mats

Weapon	Percent Fragments Stopped at Cited Range			
	5 ft	10 ft	20 ft	30 ft
81-mm mortar	95	98	98-100	98-100
82-mm mortar	98	98-100	98-100	98-100
4.2-in mortar	76	82	91	98
107-mm rocket	70	79	89	96
120-mm mortar	98	98-100	98-100	98-100
122-mm rocket	—	—	70	78

Trenches, as other fighting positions, are developed progressively. They are improved by digging deeper, from a minimum of 2 feet to about 5 1/2 feet. As a general rule, deeper excavation is desired for other than fighting trenches to provide more protection or allow more headroom. Some trenches may also -require widening to accommodate more traffic, including stretchers. It is usually necessary to revet trenches that are more than 5 feet deep in any type of soil. In the deeper trenches, some engineer advice or assistance is usually necessary in providing adequate drainage. Two basic trenches are the crawl trench and the standard fighting trench.

UNIT POSITIONS

Survivability operations are required to support the deployment of units with branch-specific missions, or missions of extreme tactical importance. These units are required to deploy and remain in one location for a considerable amount of time to perform their mission. Thus, they may require substantial protective construction.

SPECIAL DESIGNS

Table 2-3 summarizes construction estimates and levels of protection for the fighting positions, bunkers, shelters, and protective walls presented in this section.

FIGHTING POSITIONS

The following two positions are designed for use by two or more individuals armed with rifles or machine guns, Although these are beyond the construction capabilities of non-engineer troops, certain construction phases can be accomplished with little or no engineer assistance. For example, while engineer assistance may be necessary to build steel frames and cut timbers for the roof of a structure, the excavation, assembly, and installation are all within the capabilities of most units. Adequate support for overhead cover is extremely important. The support system should be strong enough to safely support the roof and soil material and survive the effects of weapon detonations.

BUNKERS

Bunkers are larger fighting positions con- strutted for squad-size units who are required to remain in defensive positions for a longer period of time. They are built either above- ground or below ground and are usually made of reinforced concrete. Because of the extensive engineer effort required to build bunkers, they are usually made during strong- point construction. If time permits, bunkers are connected to other fighting or supply positions by tunnels. Prefabrication of bunker assemblies affords rapid construction and placement flexibility. Bunkers offer excellent protection against direct fire and indirect fire effects and, if properly constructed with appropriate collective protection equipment, they provide protection against chemical and biological agents.

Table 2-3: Characteristics of special design positions

Type of Position	Estimated Construction Time (man-hours)	Equipment Requirements	Direct Small Caliber Fire	Indirect Fire Blast and Fragmentation (Near-Miss)*	Indirect Fire Blast and Fragmentation (Direct Hit)	Nuclear Weapons**	Remarks
FIGHTING POSITIONS							
Wood-frame or steel-frame fighting position with 2½-ft overhead cover	32	Hand tools	12.7mm	Medium artillery no closer than 30 ft	Small mortar	Good	
Fabric-covered frame fighting position with 1½-ft overhead cover	16	Hand tools	12.7mm	Medium artillery no closer than 15 ft	Small mortar	Good	

Note: Chemical protection is assumed because of individual protective masks and clothing.

* Shell sizes are: Small   Medium
    Mortar 82mm   120mm
    Artillery 105mm 152mm

** Nuclear protection ratings are rated poor, fair, good, very good, and excellent.

(continued)

Table 2-3: *(Continued)*

Type of Position	Estimated Construction Time (man-hours)	Equipment Requirements	Direct Small Caliber Fire	Indirect Fire Blast and Fragmentation (Near-Miss)*	Indirect Fire Blast and Fragmentation (Direct Hit)	Nuclear Weapons**	Remarks
BUNKERS							
Corrugated metal fighting bunker with 2½-ft overhead cover	48	Hand tools, backhoe	7.62mm	Medium artillery no closer than 10 ft	Small mortar	Good	
Plywood perimeter bunker	48	Hand tools, backhoe	7.62mm	Limited protection - no overhead protection	None	Poor	
Concrete log bunker with 2½-ft overhead cover	42	Hand tools, backhoe	7.62mm	Medium artillery no closer than 10 ft	Small mortar	Good	Construction time assumes precast logs. Protection provided includes one layer of sandbags around walls
Precast concrete slab bunker with 2½-ft overhead cover	30	Hand tools, backhoe, crane	7.62mm	Medium artillery no closer than 10 ft	Small mortar	Good	Construction time assumes prefabricated slabs. Protection provided includes one layer of sandbags around walls
Concrete arch bunker with 2½-ft overhead cover	38	Hand tools, backhoe, crane	7.62mm	Medium artillery no closer than 10 ft	Small mortar	Good	Construction time assumes prefabricated sections. Protection provided includes one layer of sandbags around walls
SHELTERS							
Two-soldier sleeping shelter with 2-ft overhead cover	10	Hand tools	7.62mm	Small mortar on contact	Small mortar	Fair	
Metal culvert shelter with 2-ft overhead cover	48	Hand tools, backhoe	7.62mm	Small mortar no closer than 5 ft	None	Fair	
Inverted metal shipping container shelter with 2-ft overhead cover	28	Hand tools, backhoe	12.7mm	Medium artillery no closer than 10 ft	Small mortar	Good	

Note: Chemical protection is assumed because of individual protective masks and clothing.

* Shell sizes are:

	Small	Medium
Mortar	82mm	120mm
Artillery	105mm	152mm

** Nuclear protection ratings are rated poor, fair, good, very good, and excellent.

(continued)

Table 2-3: *(Continued)*

Type of Position	Estimated Construction Time (man-hours)	Equipment Requirements	Direct Small Caliber Fire	Indirect Fire Blast and Fragmentation (Near-Miss)*	Indirect Fire Blast and Fragmentation (Direct Hit)	Nuclear Weapons**	Remarks
SHELTERS (Continued)							
Airtransportable assault with 2-ft overhead cover	60	Hand tools, backhoe	Cannot engage	Medium artillery no closer than 30 ft	Small mortar	Very good	Construction time assumes prefabricated walls and floor
Timber post buried shelter with 2½-ft overhead cover	48	Hand tools, backhoe	Cannot engage	Medium artillery no closer than 30 ft	Small mortar	Very good	
Modular timber frame shelter with 2-ft overhead cover	96	Hand tools, backhoe	Cannot engage	Medium artillery no closer than 20 ft	Small mortar	Very good	
Timber frame buried shelter with 2-ft overhead cover	84	Hand tools, backhoe	Cannot engage	Medium artillery no closer than 25 ft	Small mortar	Very good	
Aboveground cavity wall shelter with 2-ft overhead cover	700	Hand tools, backhoe, crane	12.7mm	Medium artillery no closer than 10 ft	Small mortar	Good	
Steel-frame/fabric-covered shelter with 1½-ft overhead cover	35	Hand tools, backhoe	Cannot engage	Medium artillery no closer than 10 ft	Small mortar	Very good	Construction time assumes prefabricated frame
Hardened frame/fabric shelter with 4-ft overhead cover	45	Hand tools, backhoe	Cannot engage	Medium artillery no closer than 10 ft	Medium artillery	Excellent	Shelter provides improved nuclear protection to 30 psi
Rectangular fabric/frame shelter with 1½-ft overhead cover	38	Hand tools, backhoe	Cannot engage	Medium artillery no closer than 15 ft	Medium artillery	Very good	Construction time assumes prefabricated frame
Concrete arch shelter with 4-ft overhead cover	64	Hand tools, dozer, backhoe, crane	Cannot engage	Medium artillery no closer than 5 ft	Medium artillery	Very good	Construction time assumes prefabricated arches and end walls

Note: Chemical protection is assumed because of individual protective masks and clothing.

* Shell sizes are:

	Small	Medium
Mortar	82mm	120mm
Artillery	105mm	152mm

** Nuclear protection ratings are rated poor, fair, good, very good, and excellent.

(continued)

Table 2-3: *(Continued)*

Type of Position	Estimated Construction Time (man-hours)	Equipment Requirements	Direct Small Caliber Fire	Indirect Fire Blast and Fragmentation (Near-Miss)*	Indirect Fire Blast and Fragmentation (Direct Hit)	Nuclear Weapons**	Remarks
SHELTERS (Continued)							
Metal pipe arch shelter with 4-ft overhead cover	58	Hand tools, dozer, backhoe, crane	Cannot engage	Medium artillery no closer than 5 ft	Medium artillery	Very good	Construction time assumes pre-assembled arch and end section

Type of Position	Estimated Construction Time (man-hours) per 10-ft section	Equipment Requirements	Direct Small Caliber Fire	Indirect Fire Blast and Fragmentation (Near-Miss)*	Direct Fire HEAT	Nuclear Weapons**	Remarks
PROTECTIVE WALLS							
Earth wall	3	Dozer; dump truck; scoop loader	12.7mm	Medium artillery no closer than 5 ft	120mm at wall base	Poor	
Earth wall with revetment	20	Hand tools; scoop loader	12.7mm	Medium artillery no closer than 5 ft	120mm at wall base	Poor	
Soil-cement wall	25	Hand tools; concrete mixer; crane w/concrete bucket	12.7mm	Small artillery no closer than 5 ft	82mm at wall base	Poor	Walls require forming
Soil bin wall with log revetment	35	Hand tools; scoop loader	5.45mm	Small artillery no closer than 5 ft	None	Poor	
Soil bin wall with timber revetment	30	Hand tools; scoop loader	5.45mm	Small artillery no closer than 5 ft	None	Poor	
Soil bin wall with plywood revetment	19	Hand tools; scoop loader	12.7mm	Medium artillery no closer than 5 ft	120mm at wall base	Poor	Based on plywood design. Provides nuclear blast protection for drag sensitive targets
Plywood portable wall	5	Hand tools; backhoe	5.45mm	Small mortar no closer than 5 ft	None	Poor	
Steel landing mat wall	3	Welding; crane	None	Refer to the table on page	None	Poor	M8A1 steel landing mat only

Note: Chemical protection is assumed because of individual protection masks and clothing. All walls are 5 feet high with minimum thickness as specified in construction plans.

* Shell sizes are:

	Small	Medium
Mortar	82mm	120mm
Artillery	105mm	152mm

** Nuclear protection is minimal except as noted.

(continued)

Table 2-3: *(Continued)*

Type of Position	Estimated Construction Time (man-hours) per 10-ft section	Equipment Requirements	Direct Small Caliber Fire	Indirect Fire Blast and Fragmentation (Near-Miss)*	Direct Fire HEAT	Nuclear Weapons**	Remarks
PROTECTIVE WALLS *(Continued)*							
Portable precast concrete wall	29	Hand tools; concrete mixer; crane	7.62mm	Medium artillery no closer than 5 ft	None	Poor	One layer of sandbags on outer panel surface improves small caliber protection
Cast-in-place concrete wall	35	Hand tools; concrete mixer; crane w/concrete bucket	12.7mm	Small artillery no closer than 5 ft	None	Poor	One layer of sandbags on outer panel surface improves protection to include indirect fire blast and fragmentation from large artillery
Portable asphalt armor panels 2x8x4	15	Hand tools; welding; hot asphalt source	7.62mm	Small artillery no closer than 5 ft	None	Poor	

Note: Chemical protection is assumed because of individual protection masks and clothing. All walls are 5 feet high with minimum thickness as specified in construction plans.

* Shell sizes are: Small   Medium
                    Mortar 82mm   120mm
                    Artillery 105mm 152mm

** Nuclear protection is minimal except as noted.

SHELTERS

Shelters are primarily constructed to protect soldiers, equipment, and supplies from enemy action and the weather. Shelters differ from fighting positions because there are usually no provisions for firing weapons from them. However, they are usually constructed near— or to supplement—fighting positions. When available, natural shelters such as caves, mines, or tunnels are used instead of constructing shelters. Engineers are consulted to determine suitability of caves and tunnels.

The best shelter is usually one that provides the most protection but requires the least amount of effort to construct. Shelters are frequently prepared by support troops, troops making a temporary halt due to inclement weather, and units in bivouacs, assembly areas, and rest areas. Shelters are constructed with as much overhead cover as possible. They are dispersed and limited to a maximum capacity of about 25 soldiers. Supply shelters are of any size, depending on location, time, and materials available. Large shelters require additional camouflaged entrances and exits.

All three types of shelters—below ground, aboveground, and cut-and-cover—are usually sited on reverse slopes, in woods, or in some form of natural defilade such as ravines, valleys, wadis, and other hollows or depressions in the terrain. They are not constructed in paths of natural drainage lines. All shelters require camouflage or concealment. As time permits, shelters are continuously improved.

Below ground shelters require the most construction effort but generally provide the highest level of protection from conventional, nuclear, and chemical weapons.

Cut-and-cover shelters are partially dug into the ground and backfilled on top with as thick a layer of cover material as possible. These shelters provide excellent protection from the weather and enemy action.

Above-ground shelters provide the best observation and are easier to enter and exit than below ground shelters. They also require the least amount of labor to construct, but are hard to conceal and require a large amount of cover and revetting material. They provide the least amount of protection from nuclear and conventional weapons; however, they do provide protection against liquid droplets of chemical agents. Aboveground shelters are seldom used for personnel in forward combat positions unless the shelters are concealed in woods, on reverse slopes, or among buildings. Aboveground shelters are used when water levels are close to the ground surface or when the ground is so hard that digging a below ground shelter is impractical.

The following shelters are suitable for a variety of uses where troops and their equipment require protection, whether performing their duties or resting.

PROTECTIVE WALLS

Several basic types of walls are constructed to satisfy various weather, topographical, tactical, and other military requirements. The walls range from simple ones, constructed with hand tools, to more difficult walls requiring specialized engineering and equipment capabilities.

Protection provided by the walls is restricted to stopping fragment and blast effects from near-miss explosions of mortar, rocket, or artillery shells; some direct fire protection is also provided. Overhead cover is not practical due to the size of the position surrounded by the walls. In some cases, modification of the designs shown will increase nuclear protection. The wall's effectiveness substantially increases by locating it in adequately-defended areas. The walls need close integration with other forms of protection such as dispersion, concealment, and adjacent fighting positions. The protective walls should have the minimum inside area required to perform operational duties. Further, the walls should have their height as near to the height of the equipment as practical.

CHAPTER 3

Special Operations and Situations

The two basic operations involving U.S. force deployment are combined and contingency. Combined operations are enacted in areas where U.S. forces are already established, such as NATO nations. Where few or no U.S. installations exist, usually in undeveloped regions, contingency operations are planned. In both cases, survivability missions will require intensive engineer support in all types of terrain and climate. Each environment's advantages and disadvantages are adapted to survivability planning, designing, and constructing positions. Fighting and protective positions in jungles, mountainous areas, deserts, cold regions, and urban areas require specialized knowledge, skills, techniques, and equipment. This chapter presents characteristics of five environments which impact on survivability and describes the conditions expected during combined and contingency operations.

SPECIAL TERRAIN ENVIRONMENTS

JUNGLES

Jungles are humid, tropic areas with a dense growth of trees and vegetation. Visibility is typically less than 100 feet, and areas are sparsely populated. Because mounted infantry and armor operations are limited in jungle areas, individual and crew-served weapons fighting position construction and use receive additional emphasis. While jungle vegetation provides excellent concealment from air and ground observation, fields of fire are difficult to establish. Vegetation does not provide adequate cover from small caliber direct fire and artillery indirect fire fragments. Adequate cover is available, though, if positions are located using the natural ravines and gullies produced by erosion from the area's high annual rainfall.

The few natural or locally-procurable materials which are available in jungle areas are usually limited to camouflage use. Position construction materials are transported to these areas and are required to be weather and rot resistant. When shelters are constructed in jungles, primary consideration is given to drainage provisions. Because of high amounts of rainfall and poor soil drainage, positions are built to allow for good, natural drainage routes. This technique not only prevents flooded positions but, because of nuclear fallout washing down from trees and vegetation, it also prevents positions from becoming radiation hot spots.

Other considerations are high water tables, dense undergrowth, and tree roots, often requiring above-ground level protective construction. A structure used in areas where groundwater is high, or where there is a low-pressure resistance soil, is the fighting position platform, depicted below. This platform provides a floating base or

Special Operations and Situations   83

floor where wet or low-pressure resistance soil precludes standing or sitting. The platform is constructed of small branches or timber layered over cross-posts, thus distributing the floor load over a wider area. As shown in the following two illustrations, satisfactory rain shelters are quickly constructed using easily-procurable materials such as ponchos or natural materials.

MOUNTAINOUS AREAS
Characteristics of mountain ranges include rugged, poorly trafficable terrain, steep slopes, and altitudes greater than 1,600 feet. Irregular mountain terrain provides numerous places for cover and concealment. Because of rocky ground, it is difficult and often impossible to dig below ground positions; therefore, boulders and loose rocks are used in aboveground construction. Irregular fields of fire and dead spaces are considered when designing and locating fighting positions in mountainous areas.

Reverse slope positions are rarely used in mountainous terrain; crest and near-crest positions on high ground are much more common. Direct fire weapon positions in mountainous areas are usually poorly concealed by large fields of fire. Indirect fire weapon positions are better protected from both direct and indirect fire when located behind steep slopes and ridges.

Another important design consideration in mountain terrain is the requirement for substantial overhead cover. The adverse effects of artillery bursts above a protective position are greatly enhanced by rock and gravel displacement or avalanche. Construction materials used for both structural and shielding components are most often indigenous rocks, boulders, and rocky soil. Often, rock formations are used as structural wall components without modification. Conventional tools are inadequate for preparing individual and crew-served weapons fighting positions in rocky terrain. Engineers assist with light equipment and tools (such as pneumatic jackhammers) delivered to mountain areas by helicopter.

In areas with rocky soil or gravel, wire cages or gabions are used as building blocks in protective walls, structural walls, and fighting positions. Gabions are constructed of lumber, plywood, wire fence, or any suitable material that forms a stackable container for soil or gravel.

The two-soldier mountain shelter is basically a hole 7 feet long, 3 1/2 feet wide, and 3 1/2 feet deep. The hole is covered with 6- to 8-inch diameter logs with evergreen branches, a shelter half, or local material such as topsoil, leaves, snow, and twigs placed on top. The floor is usually covered with evergreen twigs, a shelter half, or other expedient material. Entrances can be provided at both ends or a fire pit is sometimes dug at one end for a small fire or stove. A low earth parapet is built around the position to provide more height for the occupants.

DESERTS

Deserts are extensive, arid, arid treeless, having a severe lack of rainfall and extreme daily temperature fluctuations. The terrain is sandy with boulder-strewn areas, mountains, dunes, deeply-eroded valleys, areas of rock and shale, and salt marshes. Effective natural barriers are found in steep slope rock formations. Wadis and other dried up drainage features are used extensively for protective position placement.

Designers of fighting and protective positions in desert areas must consider the lack of available natural cover and concealment.

The only minimal cover available is through the use of terrain masking; therefore, positions are often completed above ground. Mountain and plateau deserts have rocky soil or "surface chalk" soil which makes digging difficult. In these areas, rocks and boulders are used for cover. Most often, parapets used in desert fighting or protective positions are undesirable because of probable enemy detection in the flat desert terrain, Deep-cut positions are also difficult to construct in soft sandy areas because of wall instability during excavations. Revetments are almost always required, unless excavations are very wide and have gently sloping sides of 45 degrees or less. Designing overhead cover is additionally important because nuclear explosions have increased fallout due to easily displaced sandy soil.

Indigenous materials are usually used in desert position construction. However, prefabricated structures and revetments for excavations, if available, are ideal. Metal culvert revetments are quickly emplaced in easily excavated sand. Sandbags and sand-filled ammunition boxes are also used for containing backsliding soil. Therefore, camouflage and concealment, as well as light and noise discipline, are important considerations during position construction. Target acquisition and observation are relatively easy in desert terrain.

COLD REGIONS

Cold regions of the world are characterized by deep snow, permafrost, seasonally frozen ground, frozen lakes and rivers, glaciers, and long periods of extremely cold temperatures. Digging in frozen or semifrozen ground is difficult with equipment, and virtually impossible for the soldier with an entrenching tool. When possible, positions are designed to take advantage of below ground cover. Positions are dug as deep as possible, then built up. Fighting and protective position construction in snow or frozen ground takes up to twice as long as positions in unfrozen ground. Also, positions used in cold regions are affected by wind and the possibility of thaw during warming periods. An unexpected thaw causes a severe drop in the soil strength which creates mud and drainage problems. Positions near bodies of water, such as lakes or rivers, are carefully located to prevent flooding damage during the spring melt season. Wind protection greatly decreases the effects of cold on both soldiers and equipment. The following areas offer good wind protection:

- Densely wooded areas.
- Groups of vegetation; small blocks of trees or shrubs.

- The lee side of terrain elevations. (The protected zone extends horizontally up to three times the height of the terrain elevation).
- Terrain depressions.

The three basic construction materials available in cold region terrain are snow, ice, and frozen soil. Positions are more effective when constructed with these three materials in conjunction with timber, stone, or other locally-available materials.

Snow

Dry snow is suitable for expedient construction than wet snow because it does not pack as well. Snow piled at road edges after clearing equipment has passed densifies and begins to harden within hours after disturbance, even at very low temperatures. Snow compacted artificially, by the wind, and after a brief thaw is even more suitable for expedient shelters and protective structures, A uniform snow cover with a minimum thickness of 10 inches is sufficient for shelter from the weather and for revetment construction. Blocks of uniform size, typically 8 by 12 by 16 inches, depending upon degree of hardness and density, are cut from the snow pack with shovels, long knives (machetes), or carpenter's saws. The best practices for

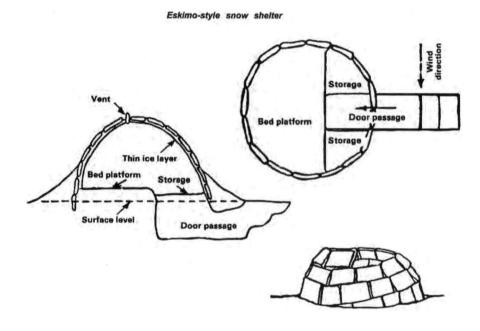

Eskimo-style snow shelter

constructing cold weather shelters are those adopted from natives of polar regions.

The systematic overlapping block-over-seam method ensures stable construction. "Caulking" seams with loose snow ensures snug, draft-free structures. Igloo shelters in cold regions have been known-survive a whole winter. An Eskimo-style snow shelter, depicted below, easily withstands above-freezing inside temperatures, thus providing comfortable protection against wind chill and low temperatures. Snow positions are built during either freezing or thawing if the thaw is not so long or intense that significant snow melt conditions occur. Mild thaw of temperatures or 2 degrees above freezing are more favorable than below-freezing temperatures because snow conglomerates readily and assumes any shape without disintegration. Below-freezing temperatures are necessary for snow construction in order to achieve solid freezing and strength. If water is available at low temperatures, expedient protective structures are built by wetting down and shaping snow, with shovels, into the desired forms.

Ice

The initial projectile-stopping capability of ice is better than snow or frozen soil; however, under sustained fire, ice rapidly cracks and collapses. Ice structures are built in the following three ways:

Layer-by-layer freezing by water. This method produces the strongest ice but, compared to the other two methods, is more time consuming. Protective surfaces are formed by spraying water in a fine mist on a structure or fabric. The most favorable temperature for this method is -10 to -15 degrees Celsius with a moderate wind. Approximately 2 to 3 inches of ice are formed per day between these temperatures (1/5-inch of ice per degree below zero).

Freezing ice fragments into layers by adding water. This method is very effective and the most frequently used for building ice structures. The ice fragments are about inch thick and prepared on nearby plots or on the nearest river or water reservoir. The fragments are packed as densely as possible into a layer 8 to 12 inches thick. Water is then sprayed over the layers of ice fragments. Crushing the ice fragments weakens the ice construction. If the weather is favorable (10 to -15 degrees Celsius with wind), a 16- to 24-inch thick ice layer is usually frozen in a day.

Laying ice blocks. This method is the quickest, but requires assets to transport the blocks from the nearest river or water reservoir to the site. Ice blocks, laid and overlapped like bricks, are of equal thickness and uniform size. To achieve good layer adhesion, the preceding layer is lightly sprayed with water before placing a new layer. Each new layer of blocks freezes onto the preceding layer before additional layers are placed.

Frozen Soil

Frozen soil is three to five times stronger than ice, and increases in strength with temperatures. Frozen soil has much better resistance to impact and explosion than to steadily-acting loads—an especially valuable feature for position construction purposes. Construction using frozen soil is performed as follows:

- Preparing blocks of frozen soil from a mixture of water and aggregate (icecrete).
- Laying prepared blocks of frozen soil.
- Freezing blocks of frozen soil together in layers.

Unfrozen soil from beneath the frozen layer is sometimes used to construct a position quickly before the soil freezes. Material made of gravel-sand-silt aggregate wetted to saturation and poured like portland cement concrete is also suitable for constructing positions. After freezing, the material has the properties of concrete. The construction methods used are analogous to those using ice. Fighting and protective positions in arctic areas are constructed both below ground and above ground.

Below ground positions. When the frost layer is one foot or less, fighting positions are usually constructed below ground, as shown. Snow packed 8 to 9 feet provides protection from sustained direct fire from small caliber weapons up to and including the Soviet 14.5-mm KPV machine gun, When possible, unfrozen excavated soil is used to form parapets about 2-foot thick, and snow is placed on the soil for camouflage and extra protection. For added frontal protection, the interior snow is reinforced with a log revetment least 3 inches in diameter. The outer surface is reinforced with small branches to initiate bullet tumble upon impact. Bullets slow down very rapidly in snow after they begin to tumble. The wall of logs directly in front of the position safely absorbs the slowed tumbling bullet.

Overhead cover is constructed with 3 feet of packed snow placed atop a layer of 6-inch diameter logs. This protection is adequate to stop indirect fire fragmentation. A layer of small, 2-inch diameter

Special Operations and Situations   89

Below ground fighting position in snow
Frontal protection
Overhead protection
Side and rear protection

logs is placed atop the packed snow to detonate quick fuzed shells before they become imbedded in the snow.

Aboveground positions. If the soil is frozen to a significant depth, the soldier equipped with only an entrenching tool and ax will have difficulty digging a fighting position. Under these conditions (below the tree line), snow and wood are often the only natural materials available to construct fighting positions. The fighting position is dug at least 20 inches deep, up to chest height, depending on snow conditions. Ideally, sandbags are used to revet the interior walls for added protection and to prevent cave-ins. If sandbags are not available, a lattice framework is constructed using small branches or, if time permits, a wall of 3-inch logs is built. Overhead cover, frontal protection, and side and rear parapets are built employing the same techniques described in chapter 2.

It is approximately ten times faster to build above-ground snow positions than to dig in frozen ground to obtain the same degree of protection. Fighting and protective positions constructed in cold regions are excavated with combined methods using handtools, excavation equipment, or explosives. Heavy equipment use is limited by traction and maneuverability.

Shelters
Shelters are constructed with a minimum expenditure of time and labor using available materials. They are ordinarily built on frozen

Dismounted TOW and machine gun positions in snow

A platform of plywood or timber is constructed to the rear of the frontal protection to provide a solid base from which to employ the guns. Overhead cover is usually offset from the firing position because of the difficulty of digging both the firing and protective positions together in the snow. The protective position should have at least 3 feet of packed snow as cover. The fighting position should have snow packed 8 to 9 feet thick for frontal, and at least 2 feet thick for side protection as shown. Sandbags are used to revet the interior walls for added protection and to prevent cave-ins. However, packed snow, rocks, 4-inch diameter logs, or ammunition cans filled with snow are sometimes used to complete the frontal and overhead protection, as well as side and rear parapets.

Individual fighting position in snow

Positions for individuals are constructed by placing packed snow on either side of a tree and extending the snow parapet 8 to 9 feet to the front, as illustrated. The side and rear parapets are constructed of a continuous snow mound, a minimum of 2 feet wide, and high enough to protect the soldier's head,

ground or dug in deep snow. Shelters that are completely above ground offer protection against the weather and supplement or replace tents. Shelter sites near wooded areas are most desirable because the wood conceals the glow of fires and provides fuel for cooking and heating. Tree branches extending to the ground offer some shelter for small units or individual protective positions.

Snow trench with wood revetment

In deep snow, trenches and weapon positions are excavated to the dimensions outlined in chapter 4. However, unless the snow is well packed and frozen, revetment is required. In snow too shallow to permit the required depth excavation, snow walls are usually constructed. The walls are made of compacted snow, revetted, and at least 6 1/2 feet thick. The table on page 5-12 contains snow wall construction requirements.

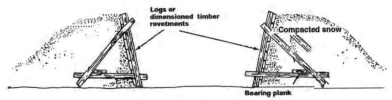

Logs or dimensioned timber revetments
Compacted snow
Bearing plank
5-6% ft between supports

Table 3-1: Snow construction for protection from grenades, small caliber fire, and HEAT projectiles

Snow Density (lb/cu ft)	Projectiles	Muzzle Velocity	Penetration, ft	Required Minimum Thickness, ft
18.0 -25.0	Grenade frag (HE)		2.0	3.0
11.2 -13.0	5.56 mm	3,250	3.8	4.4
17.4 -23.7	5.56 mm	3,250	2.3	2.6
11.2 -13.1	7.62 mm	2,750	13.0	15.0
17.4 -23.7	7.62 mm	2,750	5.2	6.0
25.5 -28.7	7.62 mm	2,750	5.0	5.8
19.9 -24.9	12.7 mm	2,910	6.4	7.4
	14.5 mm		6.0	8.0
28.1 -31.2	70 mm HEAT	900	14.0	17.5
31.2 -34.9	70 mm HEAT	900	8.7 -10.0	13.0
27.5 -34.9	90 mm HEAT	700	9.5 -11.2	14.5

Notes: These materials degrade under sustained fire. Penetrations given for 12.7 mm or smaller are for sustained fire (30 continuous firings into a 1 by 1 foot area).

Penetration characteristics of Warsaw Pact ammunitions do not differ significantly from US counterparts.

Figure given for HEAT weapons are for Soviet PRG-7 (70 mm) and United States M67 (90 mm) fired into machine-packed snow.

High explosive grenades produce small, high velocity fragments which stop in about 2 feet of packed snow. Effective protection from direct fire is independent of delivery method, including newer machine guns like the Soviet AGS-17 (30 mm) or United States MK 19/M75 (40 mm). Only armor penetrating rounds are effective.

Wigwam shelters

This shelter is constructed easily and quickly when the ground is too hard to dig and protection is required for a short bivouac. The shelter accommodates three soldiers and provides space for cooking. About 25 evergreen saplings (2 to 3 inches in diameter, 10 feet long) are cut. The limbs are left on the saplings and are leaned against a small tree so the cut ends extend about 7 feet up the trunk. The cut ends are tied together around the tree with a tent rope, wire, or other means. The ground ends of the saplings are spaced about 1 foot apart and about 7 feet from the base of the tree. The branches on the outside of the wigwam are placed flat against the saplings. Branches on the inside are trimmed off and placed on the outside to fill in the spaces. Shelter halves wrapped around the outside make the wigwam more windproof, especially after it is covered with snow. A wigwam is also constructed by lashing the cut ends of the saplings together instead of leaning them against the tree.

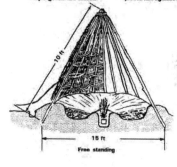

Free standing

Built around a tree

Lean-to shelter

This shelter is made of the same material as the wigwam (natural saplings woven together and brush). The saplings are placed against a rock wall, a steep hillside, a deadfall, or some other existing vertical surface, on the leeward side. The ends are closed with shelter halves or evergreen branches.

Shelter half

Snow cave

Snow caves are made by burrowing into a snowdrift and fashioning a room of the desired size. This shelter gives good protection from freezing weather and a maximum amount of concealment. The entrance slopes upward for best protection against cold air penetration. Snow caves are usually built large enough for several soldiers if the consistency of the snow prevents cave-in. Two entrances are usually used while the snow is taken out of the cave; one entrance is refilled with snow when the cave is completed. Fires in snow caves are kept small to prevent melting the structure. To allow incoming fresh air, the door is not completely sealed.

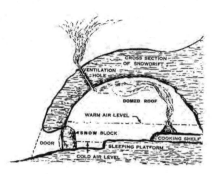

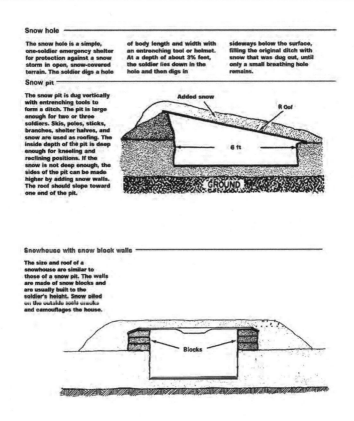

Constructing winter shelters begins immediately after the halt to keep the soldiers warm. Beds of foliage, moss, straw, boards, skis, shelter halves, and ponchos are sometimes used as protection against ground dampness and cold. The entrance to the shelter, located on the side least exposed to the wind, is close to the ground and slopes up into the shelter. Openings or cracks in the shelter walls are caulked with an earth and snow mixture to reduce wind effects. The shelter itself is constructed as low to the ground as possible. Any fire built within the shelter is placed low in fire holes and cooking pits. Although snow is windproof, a layer of insulating material, such as a shelter half or blanket, is placed between the occupant and the snow to prevent body heat from melting the snow.

URBAN AREAS

Survivability of combat forces operating in urban areas depends on the leader's ability to locate adequate fighting and protective positions from the many apparent covered and concealed areas available. Fighting and protective positions range from hasty positions formed from piles of rubble, to deliberate positions located inside urban structures. Urban structures are the most advantageous locations for individual fighting positions. Urban structures are usually divided into groups of below ground and above-ground structures.

Below Ground Structures

A detailed knowledge of the nature and location of below ground facilities and structures is of potential value when planning survivability operations in urban terrain. Typical underground street cross sections are shown in Figure 3-13.

Sewers are separated into sanitary, storm, or combined systems. Sanitary sewers carry wastes and are normally too small for troop movement or protection. Storm sewers, however, provide rainfall removal and are often large enough to permit troop and occasional vehicle movement and protection. Except for groundwater, these

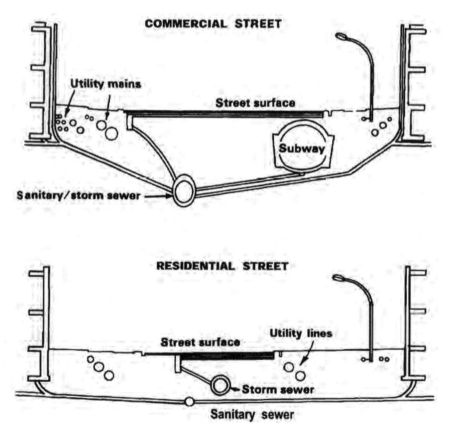

Figure 3-13: Cross sections of streets

sewers are dry during periods of no precipitation. During rainstorms, however, sewers fill rapidly and, though normally drained by electrical pumps, may overflow. During winter combat, snow melt may preclude daytime below ground operations. Another hazard is poor ventilation and the resultant toxic fume build-up that occurs in sewer tunnels and subways. The conditions in sewers provide an excellent breeding ground for disease, which demands proper troop hygiene and immunization.

Subways tend to run under main roadways and have the potential hazard of having electrified rails and power leads. Passageways often extend outward from underground malls or storage areas, and catacombs are sometimes encountered in older sections of cities.

Aboveground Structures

Aboveground structures in urban areas are generally of two types: frameless and framed.

Frameless structures. In frameless structures, the mass of the exterior wall performs the principal load-bearing functions of supporting dead weight of roofs, floors, ceilings; weight of furnishings and occupants; and horizontal loads. Frameless structures are shown in Figure 3-14.

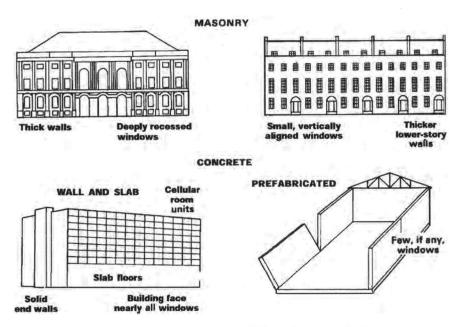

Figure 3-14: Frameless building characteristics

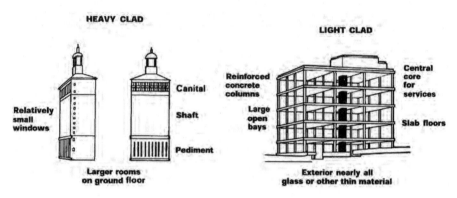

Figure 3-15: Framed building characteristics

Building materials for frameless structures include mud, stone, brick, cement building blocks, and reinforced concrete. Wall thickness varies with material and building height. Frameless structures have thicker walls than framed structures, and therefore are more resistant to projectile penetration. Fighting from frameless buildings is usually restricted to the door and window areas.

Frameless buildings vary with function, age, and cost of building materials. Older institutional buildings, such as churches, are frequently made of stone. Reinforced concrete is the principal material for wall and slab structures (apartments and hotels) and for prefabricated structures used for commercial and industrial purposes. Brick structures, the most common type of buildings, dominate the core of urban areas (except in the relatively few parts of the world where wood-framed houses are common). Close-set brick structures up to five stories high are located on relatively narrow streets and form a hard, shock-absorbing protective zone for the inner city. The volume of rubble produced by their full or partial demolition provides countless fighting positions.

Framed structures. Framed structures typically have a skeletal structure of columns and beams which supports both vertical and horizontal loads. Exterior (curtain) walls are nonload bearing. Without the impediment of load bearing walls, large open interior spaces offer little protection. The only available refuge is the central core of reinforced concrete present in many of these buildings (for example, the elevator shaft). Multistoried steel and concrete-framed structures occupy the valuable core area of most modern cities. Examples of framed structures are shown in Figure 3-15.

Material and Structural Characteristics

Urban structures, frameless and framed, fit certain material generalities. Table 3-2 converts building type and material into height/wall thicknesses. Most worldwide urban areas have more than 60 percent of their construction formed from bricks. The relationship between building height and thickness of the average brick wall is shown in Table 3-3.

Table 3-2: Urban structure material thicknesses

Building Material	Height (stories)	Average Wall Thickness, in
Frameless Structures		
Stone	1-10	30
Brick	1-3	9
Brick	3-6	15
Concrete block	1-5	8
Concrete, wall and slab	1-10	9-15
Concrete, prefabricated	1-3	7
Framed Structures		
Wood	1-5	1
Steel (heavy cladding)	3-100	5
Concrete/steel (light cladding)	3-50	1-3

Table 3-3: Average brick wall thickness

Height (stories)	Wall Thickness, in					
	1st	2nd	3rd	4th	5th	6th
1	11½					
2	13½	10½				
3	14½	13½	10½			
4	15½	14½	13½	11½		
5	18½	15½	14½	13½	12½	
6	18½	18½	15½	14½	13½	12½

SPECIAL URBAN AREA POSITIONS
Troop Protection
After urban structures are classified as either frameless or framed, and some of their material characteristics are defined, leaders evaluate them for protective soundness. The evaluation is based on troop protection available and weapon position employment requirements for cover, concealment, and routes of escape. Table 3-4 summarizes survivability requirements for troop protection.

Cover. The extent of building cover depends on the proportion of walls to windows. It is necessary to know the proportion of non-windowed wall space which might serve as protection. Frameless buildings, with their high proportion of walls to windows, afford more substantial cover than framed buildings having both a lower proportion of wall to window space and thinner (nonload bearing) walls.

Composition and thickness of both exterior and interior walls also have a significant bearing on cover assessment. Frameless buildings with their strong weight-bearing walls provide more cover than the curtain walls of framed buildings. However, interior walls of the older, heavy-clad, framed buildings are stronger than those of

Table 3-4: Survivability requirements for troops in urban buildings.

Requirements	Building Characteristics
Cover	1. Proportion of walls to windows
	2. Wall composition and thickness
	3. Interior wall and partition composition and thickness
	4. Stair and elevator modules
Concealment	1. Proportion of walls to windows
	2. Venting pattern
	3. Floor plan (horizontal and vertical)
	4. Stair and elevator modules (framed high-rise buildings)
Escape	1. Floor plan (horizontal and vertical)
	2. Stair and elevator modules

the new, light-clad, framed buildings. Cover within these light-clad framed buildings is very slight except in and behind their stair and elevator modules which are usually constructed of reinforced concrete. Familiarity with the location, dimension, and form of these modules is vital when assessing cover possibilities.

Concealment. Concealment considerations involve some of the same elements of building construction, but knowledge of the venting (window) pattern and floor plan is added.

These patterns vary with type of building construction and function. Older, heavy-clad framed buildings (such as office buildings) frequently have as full a venting pattern as possible, while hotels have only one window per room. In the newer, light-clad framed buildings, windows are sometimes used as a nonload bearing curtain wall. If the windows are all broken, no concealment possibilities exist. Another aspect of concealment— undetected movement within the building—depends on a knowledge of the floor plan and the traffic pattern within the building on each floor and from floor to floor.

Escape. In planning for escape routes, the floor plan, traffic patterns, and the relationships between building exits are considered. Possibilities range from small buildings with front street exits (posing unacceptable risks), to high-rise structures having exits on several floors, above and below ground level, and connecting with other buildings as well.

Fighting Positions

Survivability requirements for fighting positions for individuals, machine guns, and antitank and antiaircraft weapons are summarized in Table 3-5.

Individual fighting positions. An upper floor area of a multistoried building generally provides sufficient fields of fire, although corner windows can usually encompass more area. Protection from the possibility of return fire from the streets requires that the soldier know the composition and thickness of the building's outer wall. Load bearing walls generally offer more protection than the curtain walls of framed buildings. However, the relatively thin walls of a low brick building (only two-bricks thick or 8 inches) is sometimes less effective than a 15-inch thick nonload bearing curtain wall of a high-rise framed structure.

The individual soldier is also concerned about the amount of overhead protection available. Therefore, the soldier needs to know about the properties of roof, floor, and ceiling materials. These materials vary with the type of building construction. In brick buildings, the material for the ceiling of the top floor is far lighter than that for the next floor down that performs as both ceiling and floor, and thus is capable of holding up the room's live load.

Table 3-5: Survivability requirements for fighting positions in urban buildings

Individual positions	1. Wall composition and thickness of upper floors
	2. Roof composition and thickness
	3. Floor and ceiling composition and thickness
Machine gun positions	1. Wall composition and thickness
	2. Local terrain
Antitank weapon positions	1. Wall composition and thickness
	2. Room dimensions and volume
	3. Function related interior furnishings, and so forth
	4. Fields of fire (relative position of building)
	5. Arming distance
	6. Line-of-sight
Antiaircraft weapon positions	1. Roof composition and thickness
	2. Floor plan (horizontal and vertical)
	3. Line-of-sight

CHAPTER 4
Position Design Details

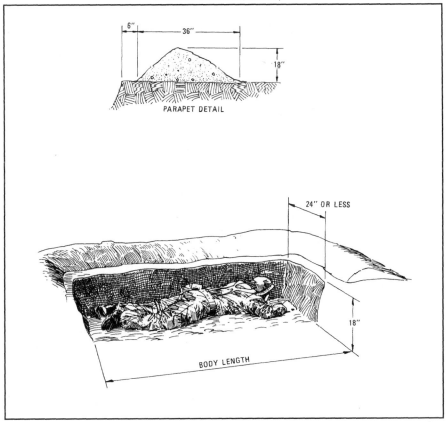

PRONE POSITION (HASTY)

ONE-SOLDIER POSITION (DELIBERATE)

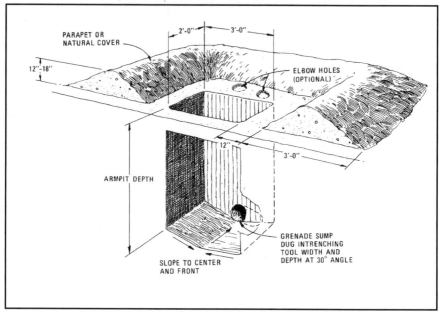

Position Design Details 103

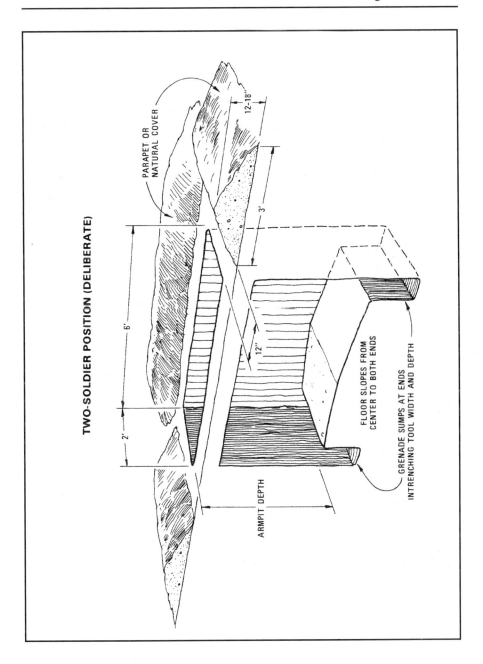

ONE- OR TWO-SOLDIER POSITION WITH OVERHEAD COVER (DELIBERATE)

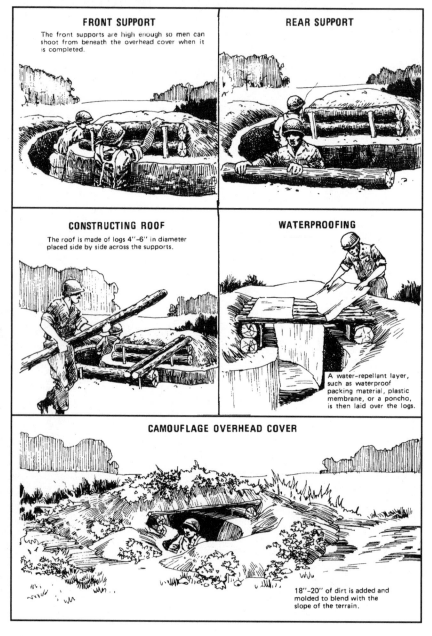

FRONT SUPPORT — The front supports are high enough so men can shoot from beneath the overhead cover when it is completed.

REAR SUPPORT

CONSTRUCTING ROOF — The roof is made of logs 4"–6" in diameter placed side by side across the supports.

WATERPROOFING — A water-repellant layer, such as waterproof packing material, plastic membrane, or a poncho, is then laid over the logs.

CAMOUFLAGE OVERHEAD COVER — 18"–20" of dirt is added and molded to blend with the slope of the terrain.

Position Design Details 105

DISMOUNTED TOW POSITION

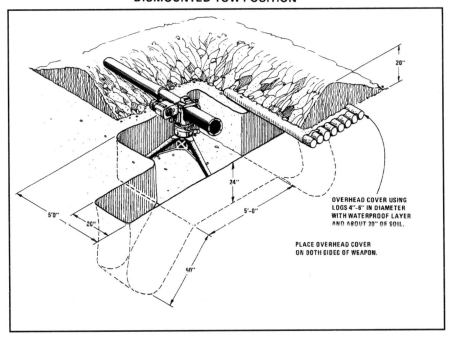

MACHINE GUN POSITION

TRACING OUTLINE

MARKING THE POSITION OF THE TRIPOD LEGS AND THE LIMITS OF THE SECTORS OF FIRE

DIGGING POSITION

THE WEAPON IS LOWERED BY DIGGING DOWN FIRING PLATFORMS WHERE THE MG WILL BE PLACED. THE CREW THEN DIGS THE HOLE ABOUT ARMPIT DEEP. DIRT IS PLACED FIRST WHERE FRONTAL COVER IS NEEDED THEN ON THE FLANKS AND REAR.

NO SECONDARY SECTOR

IN SOME POSITIONS, AN MG MAY NOT HAVE A SECONDARY SECTOR OF FIRE; SO, ONLY HALF OF THE POSITION IS DUG.

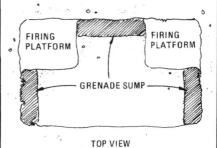

GRENADE SUMP LOCATIONS

TOP VIEW

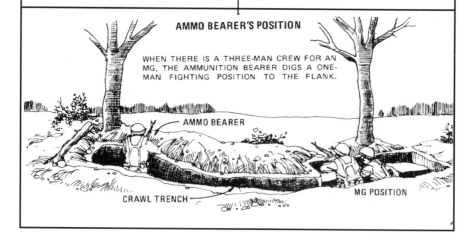

AMMO BEARER'S POSITION

WHEN THERE IS A THREE-MAN CREW FOR AN MG, THE AMMUNITION BEARER DIGS A ONE-MAN FIGHTING POSITION TO THE FLANK.

Position Design Details   107

MORTAR POSITION (81MM AND 4.2-IN MORTARS)

HASTY POSITION

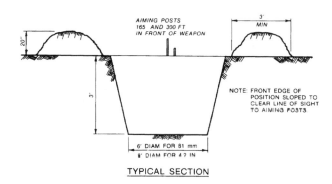

TYPICAL SECTION

IMPROVED POSITION

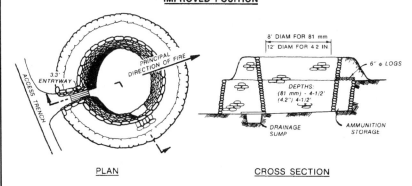

PLAN   CROSS SECTION

NOTE: TOTAL DEPTH INCLUDES THE PARAPET.
SLOPE FLOOR TOWARD DRAINAGE SLUMP

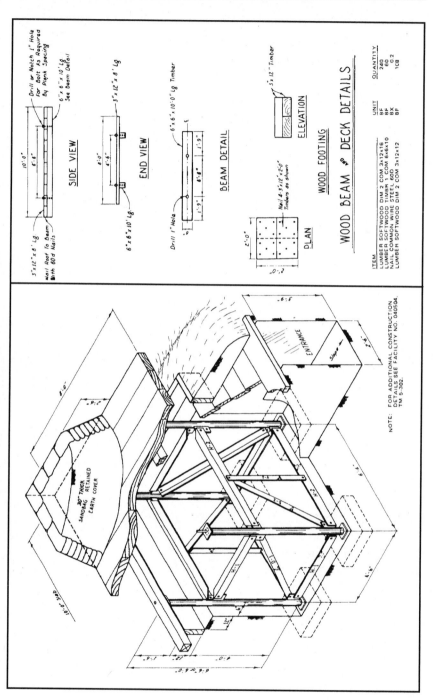

Position Design Details 109

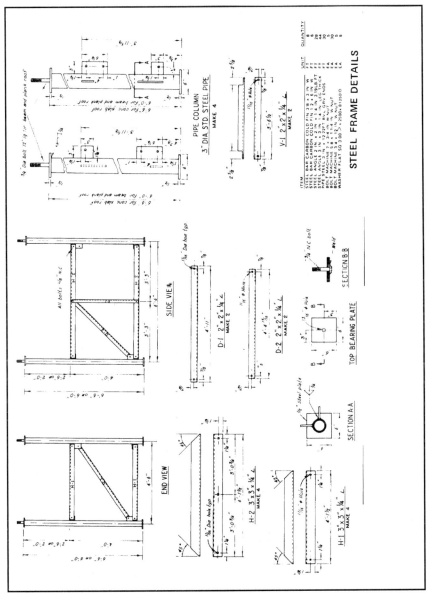

WOOD-FRAME FIGHTING POSITION (sheet 2 of 3)

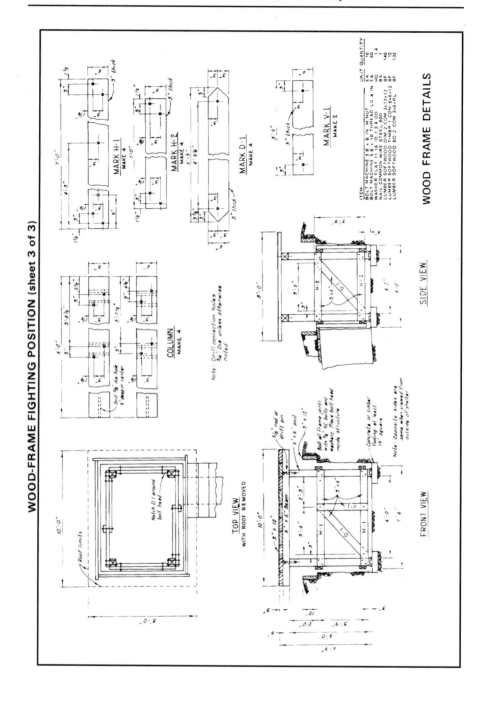

Position Design Details 111

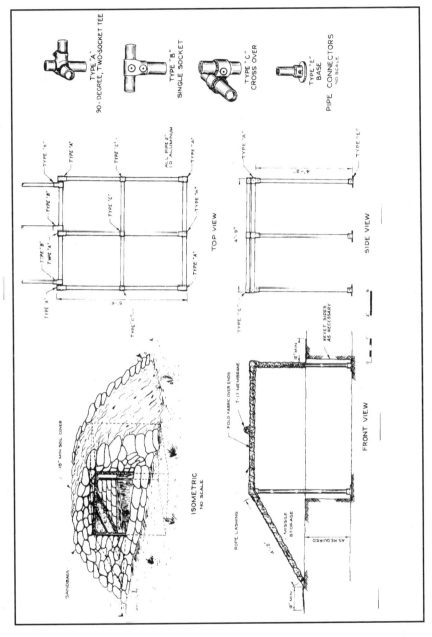

FABRIC-COVERED FRAME POSITION (sheet 1 of 2)

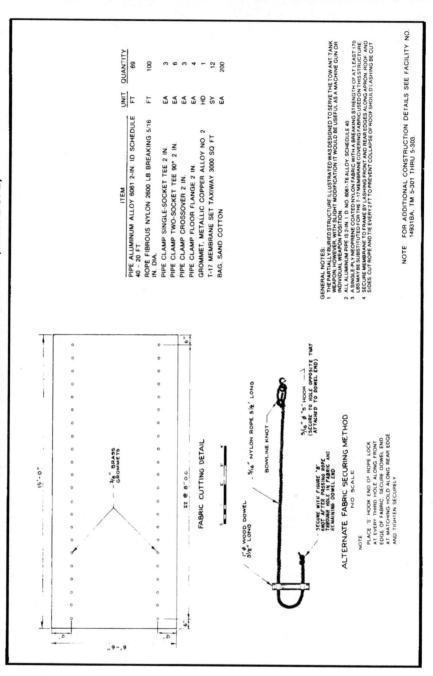

Position Design Details 113

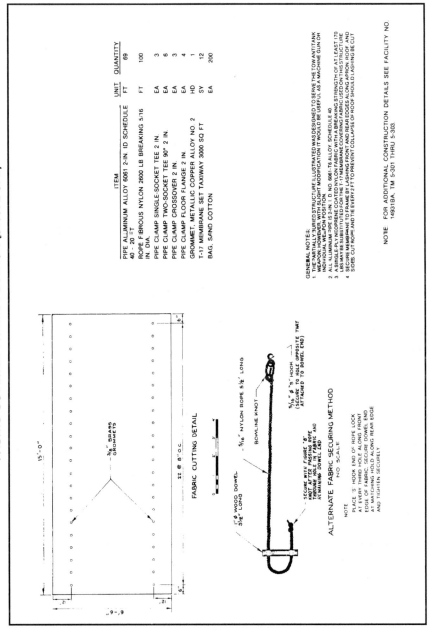

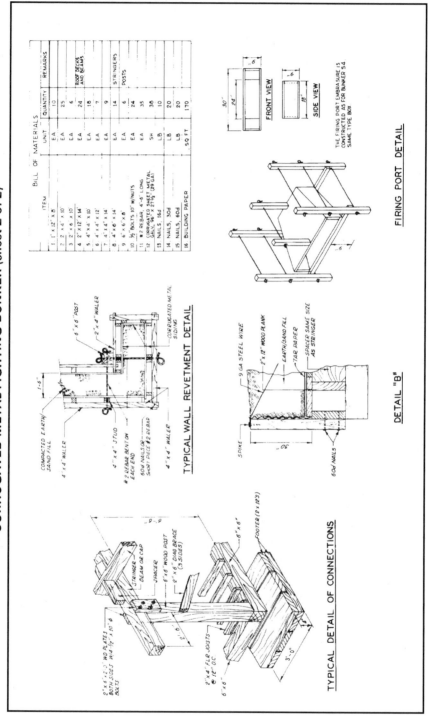

Position Design Details 115

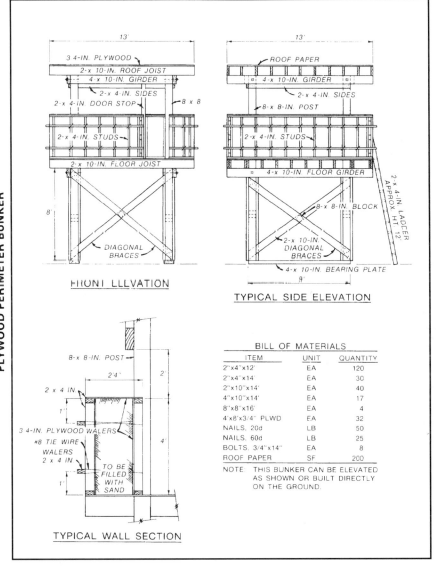

PLYWOOD PERIMETER BUNKER

FRONT ELEVATION

TYPICAL SIDE ELEVATION

TYPICAL WALL SECTION

BILL OF MATERIALS

ITEM	UNIT	QUANTITY
2"x4"x12'	EA	120
2"x4"x14'	EA	30
2"x10"x14'	EA	40
4"x10"x14'	EA	17
8"x8"x16'	EA	4
4'x8'x3/4" PLWD	EA	32
NAILS, 20d	LB	50
NAILS, 60d	LB	25
BOLTS, 3/4"x14"	EA	8
ROOF PAPER	SF	200

NOTE: THIS BUNKER CAN BE ELEVATED AS SHOWN OR BUILT DIRECTLY ON THE GROUND.

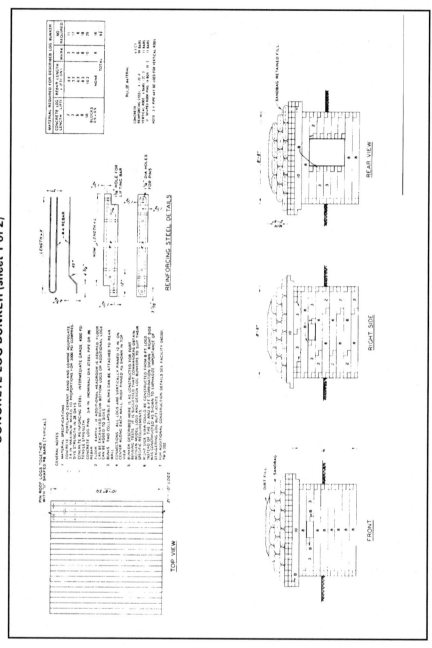

Position Design Details 117

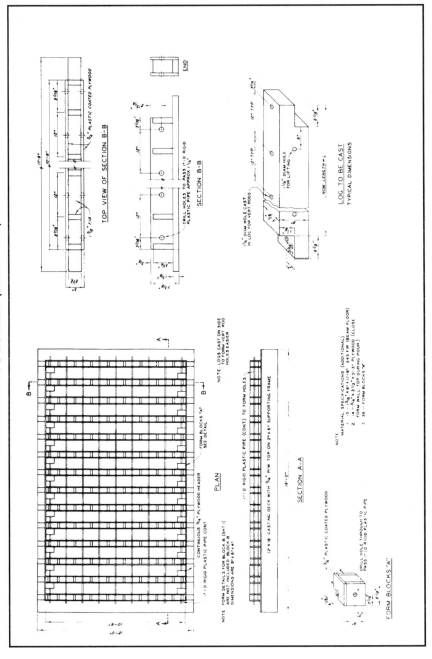

118   The Complete Guide to Shelter Skills, Tactics, and Techniques

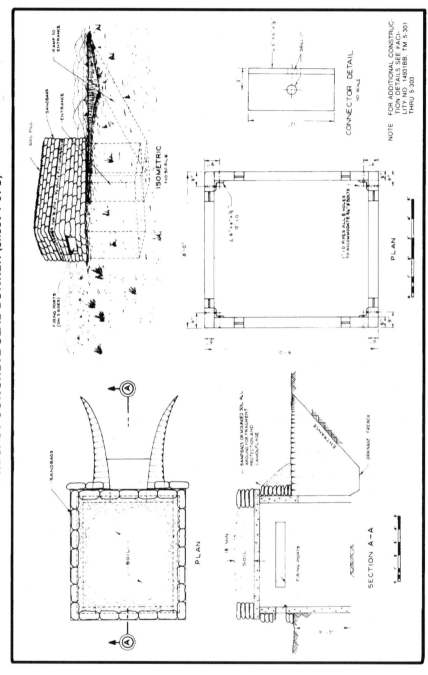

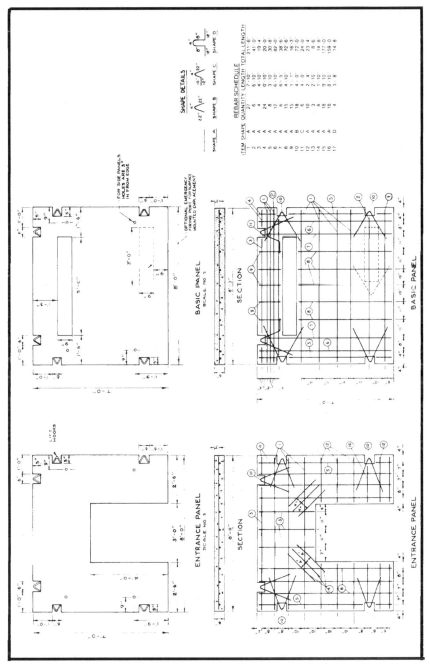

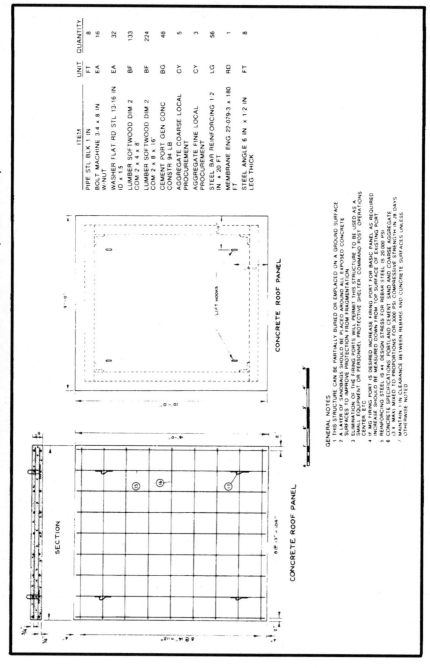

Position Design Details 121

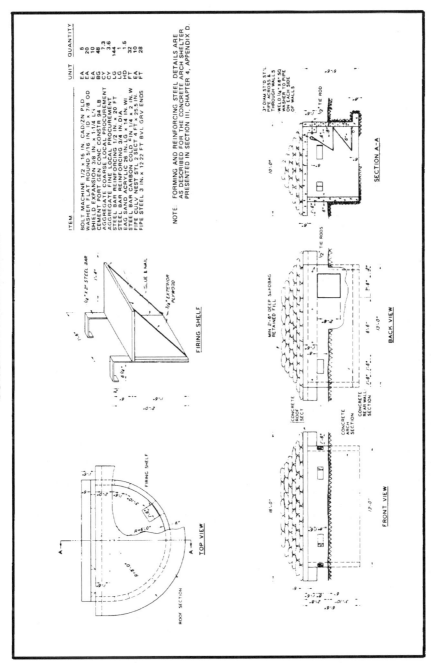

122 The Complete Guide to Shelter Skills, Tactics, and Techniques

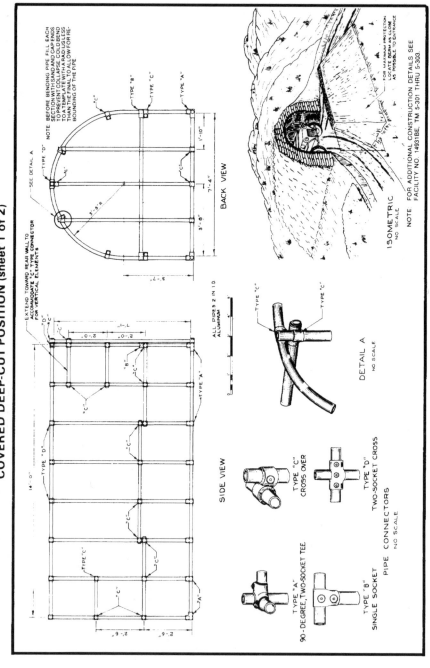

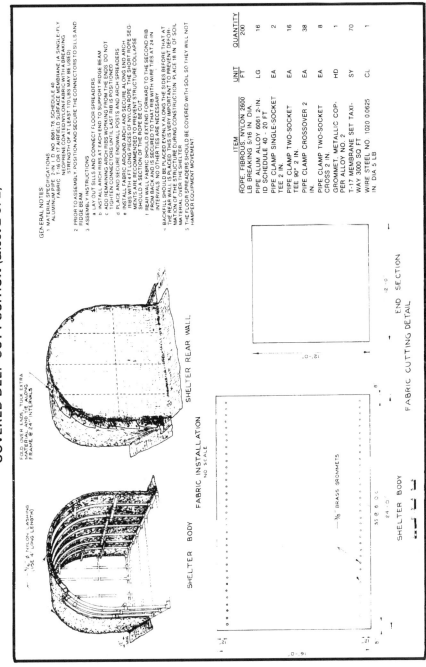

ARTILLERY FIRING PLATFORM (155MM, 175MM, AND 8-IN ARTILLERY) (sheet 1 of 3)

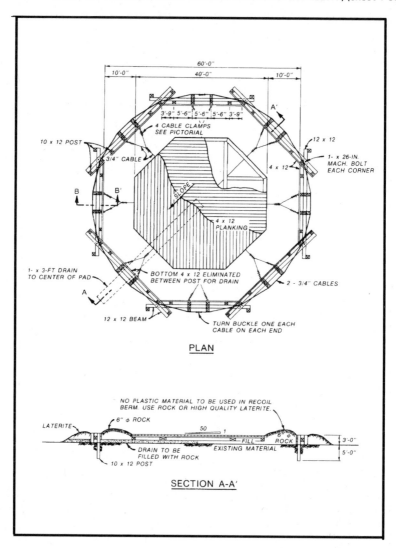

Position Design Details   125

ARTILLERY FIRING PLATFORM (155MM, 175MM, AND 8-IN ARTILLERY) (sheet 2 of 3)

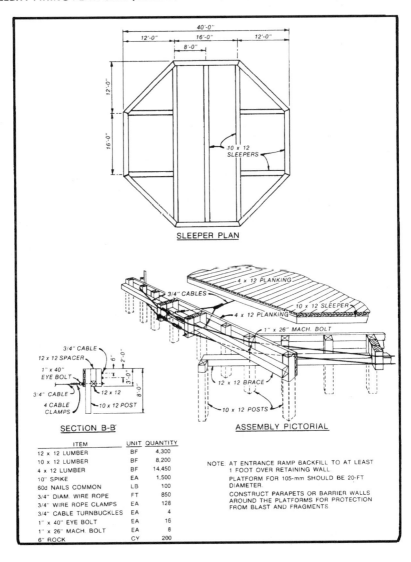

ITEM	UNIT	QUANTITY
12 x 12 LUMBER	BF	4,300
10 x 12 LUMBER	BF	8,200
4 x 12 LUMBER	BF	14,450
10" SPIKE	EA	1,500
60d NAILS COMMON	LB	100
3/4" DIAM. WIRE ROPE	FT	850
3/4" WIRE ROPE CLAMPS	EA	128
3/4" CABLE TURNBUCKLES	EA	4
1" x 40" EYE BOLT	EA	16
1" x 26" MACH. BOLT	EA	8
6" ROCK	CY	200

NOTE: AT ENTRANCE RAMP BACKFILL TO AT LEAST 1 FOOT OVER RETAINING WALL.
PLATFORM FOR 105-mm SHOULD BE 20-FT DIAMETER.
CONSTRUCT PARAPETS OR BARRIER WALLS AROUND THE PLATFORMS FOR PROTECTION FROM BLAST AND FRAGMENTS.

ARTILLERY FIRING PLATFORM (155MM, 175MM, AND 8-IN ARTILLERY) (sheet 3 of 3)

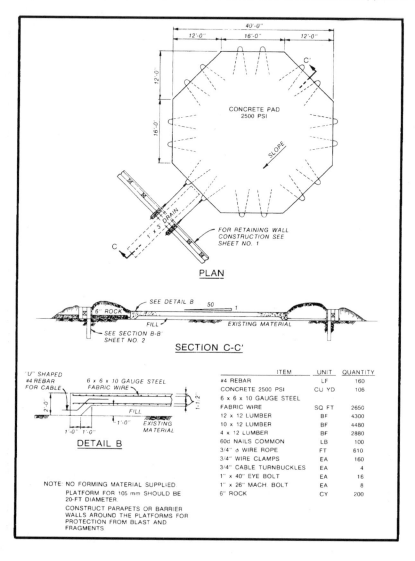

PARAPET POSITION FOR ADA

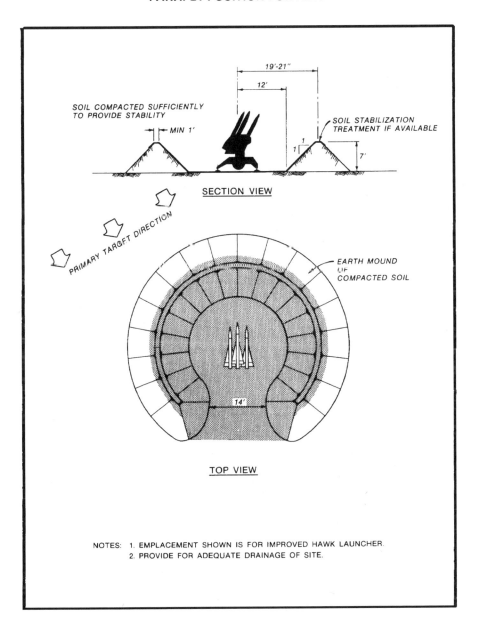

NOTES: 1. EMPLACEMENT SHOWN IS FOR IMPROVED HAWK LAUNCHER.
2. PROVIDE FOR ADEQUATE DRAINAGE OF SITE.

128 The Complete Guide to Shelter Skills, Tactics, and Techniques

TWO-SOLDIER SLEEPING SHELTER

- SCRAP BOARDS FROM AMMO BOXES AS EAVES
- CORRUGATED METAL PIPE 1/2 SECTIONS
- SANDBAGS FRONT AND BACK ONLY. FILL BETWEEN WITH SOIL
- SHALLOW EXCAVATION 8" – 10"
- 8'-0"
- 36" OR 48"

Position Design Details 129

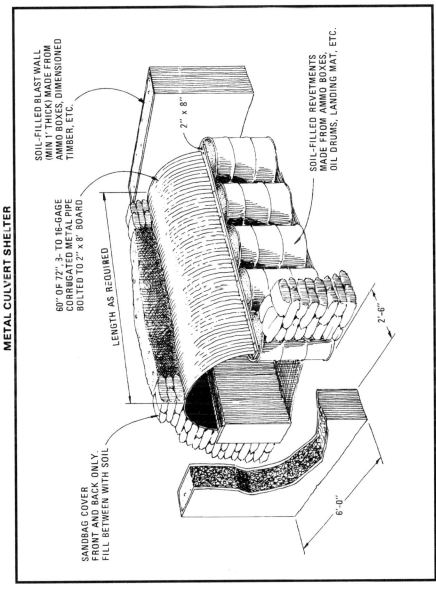

AIRTRANSPORTABLE ASSAULT SHELTER (sheet 1 of 3)

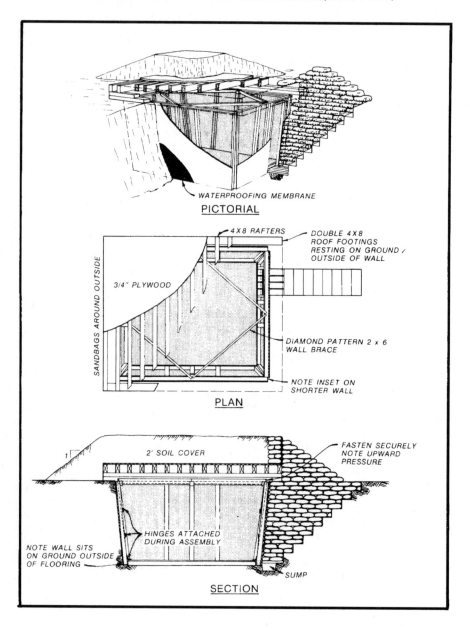

AIRTRANSPORTABLE ASSAULT SHELTER (sheet 2 of 3)

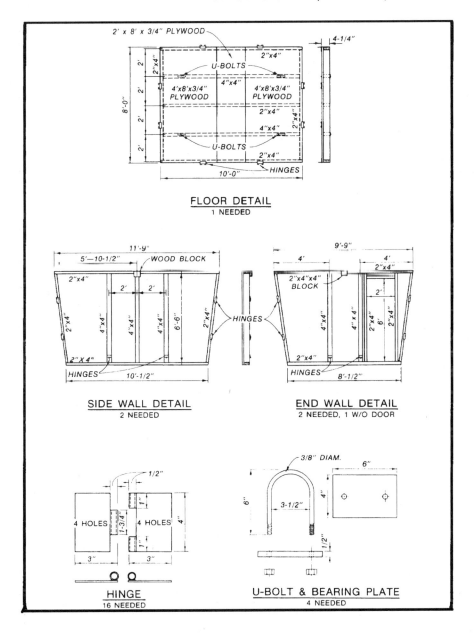

AIRTRANSPORTABLE ASSAULT SHELTER (sheet 3 of 3)

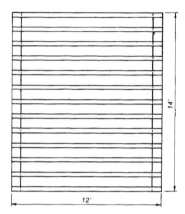

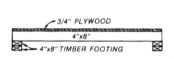

ROOF DETAIL

BILL OF MATERIALS (WALLS AND FLOOR)		
ITEM	UNITS	QUANTITY
4'x8'x3/4" PLYWOOD	EA	14
4"x4"x8'	EA	10
4"x4"x10'	EA	2
2"x4"x12'	EA	4
2"x4"x10'	EA	9
2"x4"x8'	EA	10
2"x6"x10'	EA	4
TRIM (METAL EDGING) OPTIONAL	FT	190
BOLTS (FOR HINGES)	EA	128
WOOD SCREWS (OR #8 NAILS)	LB	5
PAINT	GAL	1
HINGES	EA	16
U-BOLTS W/ BEARING PLATES	EA	4

BILL OF MATERIALS (ROOF)		
ITEM	UNIT	QUANTITY
4"x8"x12'	EA	13
4"x8"x14'	EA	4
4'x8'x3/4" PLYWOOD	EA	6

NOTES:

(1) Abut longer side walls against shorter end walls because the longer walls must sustain the greatest load. The shorter walls then act as a support. Install hinges during assembly.

(2) Provide wall bracing (2" x 6") at the top of the shelter. Brace from the center of each wall to the center of each adjacent wall (diamond pattern).

(3) Attach a sheet of plastic or other thin waterproof covering around the outside before backfilling to minimize friction between earth and the walls and increase moisture resistance.

(4) Make the shelter no larger than necessary. It should be no more than 6-1/2 feet high and the floor area should be less than 100 ft^2 unless special effort is made to provide adequate structural members in addition to those specified.

(5) Backfilling should be accomplished by hand labor, maintaining a uniform load around the perimeter as backfilling progresses.

(6) Make the bottom of the excavation 2 feet longer and 2 feet wider than the length and width of the structure floor to increase working room during erection and provide adequate clearance for the walls.

(7) Use explosives as extensively as practical during excavation to minimize required hand digging.

(8) To complete the structure provide a suitable entryway. Drainage ditches should be provided around the shelter to carry away runoff, and a waterproof cover placed over the overhead cover to prevent saturation of the soil material and eliminate seepage into the interior.

(9) Prior to lifting the structure from the installed position, remove some of the backfill with hand tools to reduce effects of wall friction.

Position Design Details 133

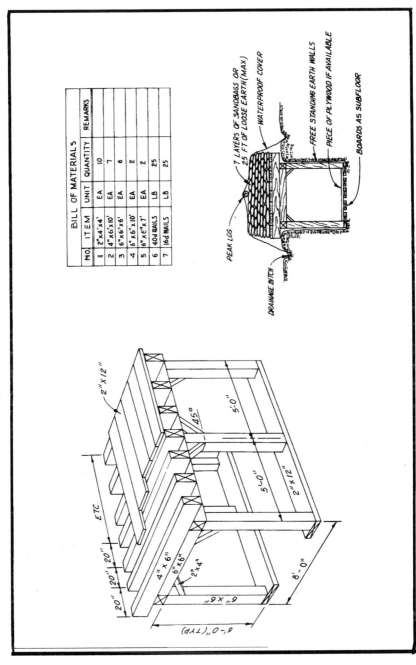

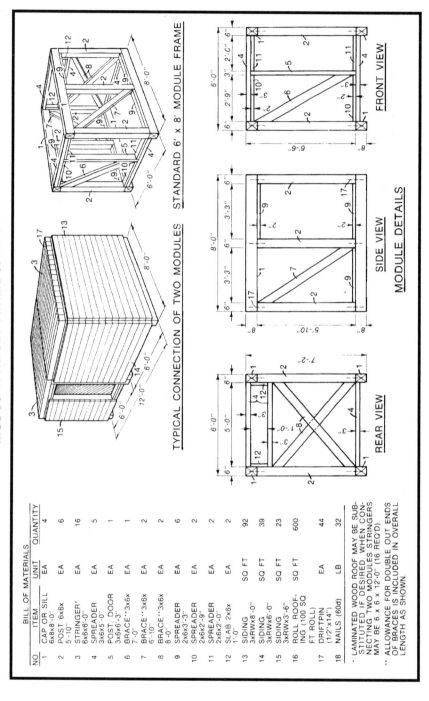

Position Design Details 135

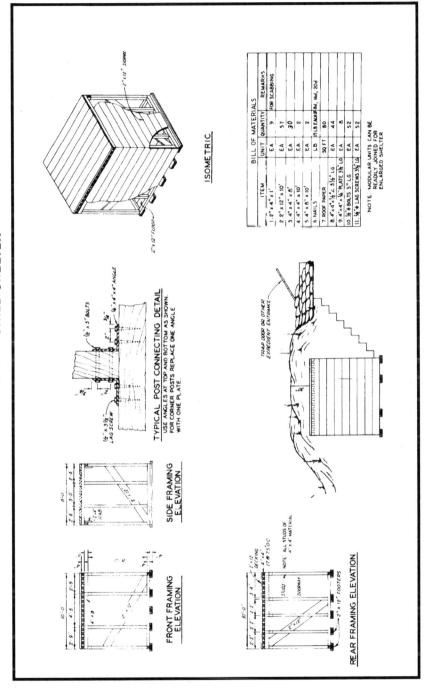

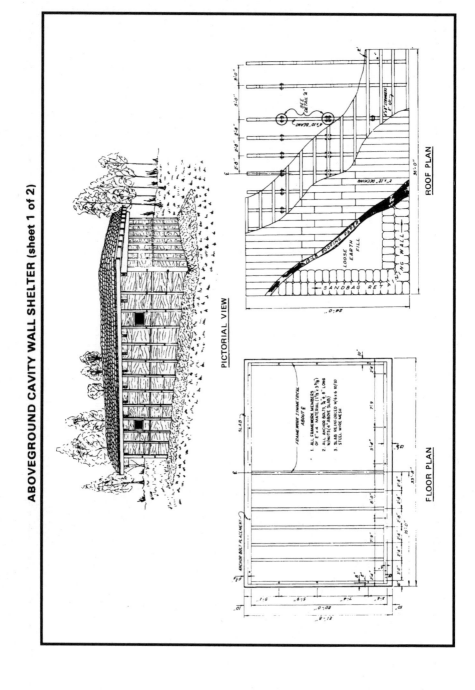

Position Design Details 137

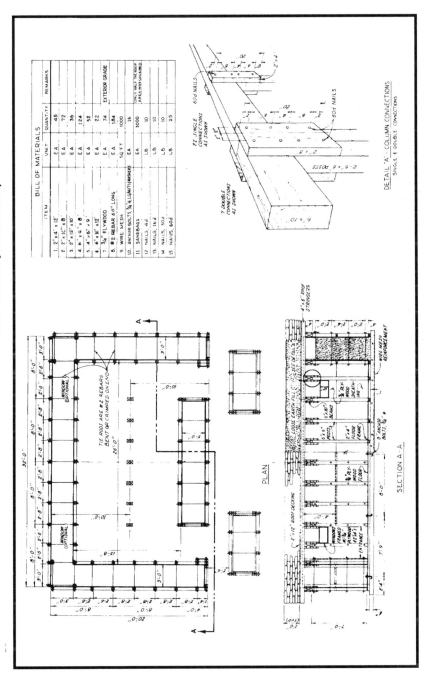

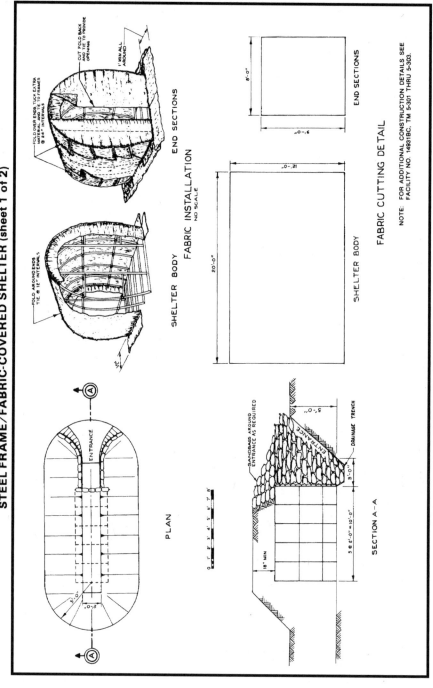

Position Design Details 139

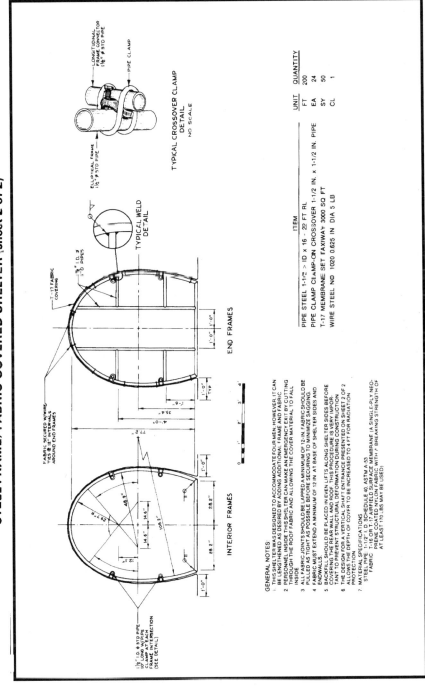

140 The Complete Guide to Shelter Skills, Tactics, and Techniques

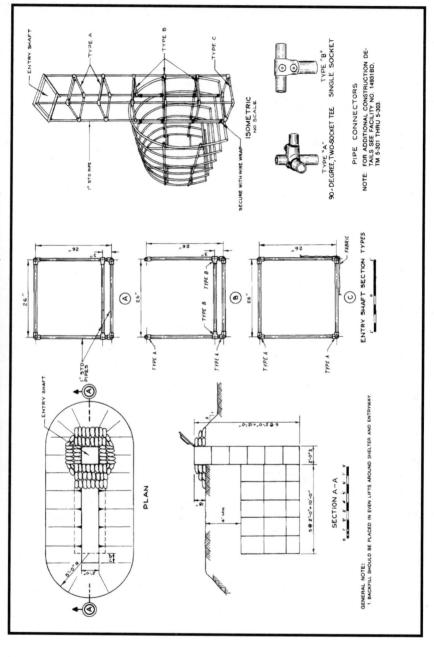

HARDENED FRAME/FABRIC SHELTER (sheet 1 of 3)

Position Design Details 141

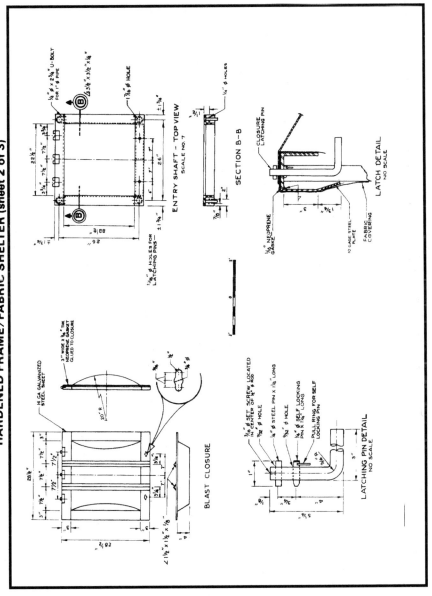

HARDENED FRAME/FABRIC SHELTER (sheet 3 of 3)

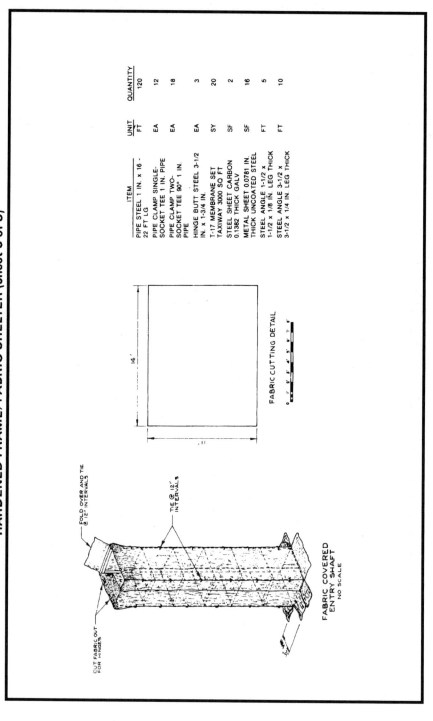

ITEM	UNIT	QUANTITY
PIPE STEEL 1 IN. x 16 - 22 FT LG	FT	120
PIPE CLAMP SINGLE-SOCKET TEE 1 IN. PIPE	EA	12
PIPE CLAMP TWO-SOCKET TEE 90° 1 IN. PIPE	EA	18
HINGE BUTT STEEL 3-1/2 IN. x 1-3/4 IN.	EA	3
T-17 MEMBRANE SET TAXIWAY 3000 SQ FT	SY	20
STEEL SHEET CARBON 0.1382 THICK GALV	SF	2
METAL SHEET 0.0781 IN. THICK UNCOATED STEEL	SF	16
STEEL ANGLE 1-1/2 x 1-1/2 x 1/8 IN. LEG THICK	FT	5
STEEL ANGLE 3-1/2 x 3-1/2 x 1/4 IN. LEG THICK	FT	10

FABRIC CUTTING DETAIL

Position Design Details 143

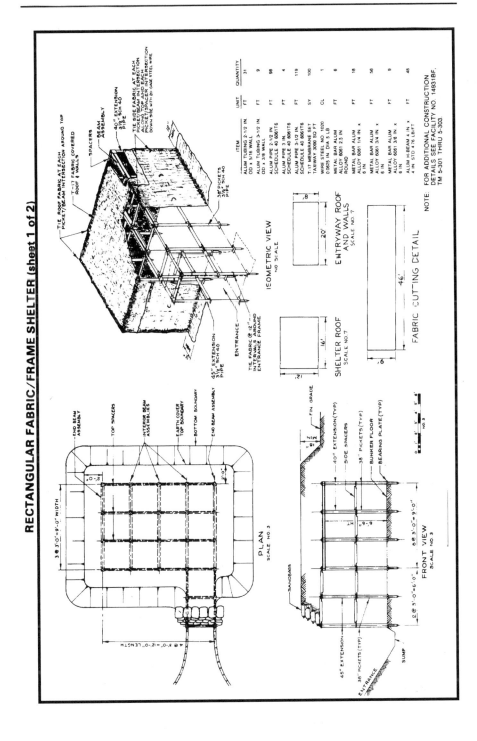

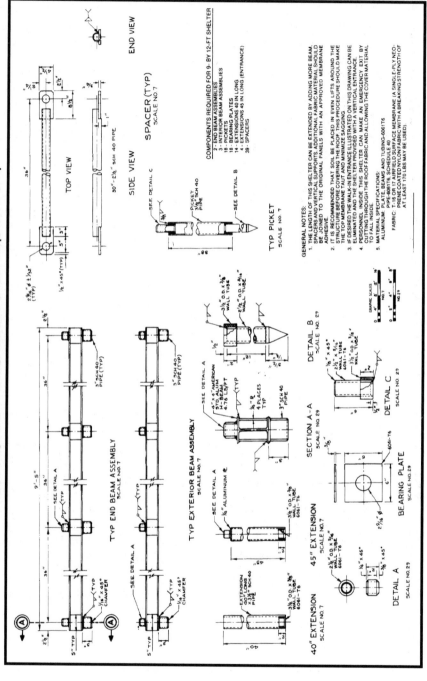

CONCRETE ARCH SHELTER (sheet 1 of 3)

GENERAL NOTES

1. MATERIAL SPECIFICATIONS
   CONCRETE: PORTLAND CEMENT, SAND & COARSE AGGREGATE 3/4" MAX, MIXED TO PROPORTIONS FOR 3000 PSI COMPRESSIVE STRENGTH IN 28 DAYS.
   CONCRETE REINFORCING STEEL: INTERMEDIATE GRADE 40,000 PSI TENSILE STRENGTH
   MULTIPLATE PIPE ARCH: 8-GAGE CORRUGATED STEEL, 6" x 2" CORRUGATIONS, GALVANIZED.
   MISCELLANEOUS FASTENERS: GRADE AS AVAILABLE.
2. DESIGN LOADING: 8' EARTH COVER AT CROWN.
3. GROUNDWATER SEEPAGE: ALL EXTERIOR SURFACES OF SHELTER & ENTRANCEWAY MUST BE COVERED WITH WATERPROOFING MEMBRANE TO PREVENT GROUND WATER SEEPAGE.
4. FLOOR: CORRUGATED-STEEL FLOOR PROVIDED.
5. OCCUPANCY: 12' x 12' BASIC SHELTER DESIGNED FOR 10 MEN ASSUMING ONE SLEEPING BUNK PER MAN.
6. EQUIPMENT PROVIDED: BASIC SHELTER PROVIDES ONLY STRUCTURAL COMPONENTS AND ASSEMBLIES. ALL EQUIPMENT MUST BE PROVIDED SEPARATELY.
7. EMERGENCY EXIT: USE ONE OF DOOR OPENINGS AS EMERGENCY EXIT.
8. BUNK INSTALLATION: TO BE SUSPENDED FROM ARCH SECTION BY ANCHOR HOOKS.
9. TRANSPORTABILITY: ARCH SECTION 4300 LB, ENDWALL 7900 LB.
10. FOR ADDITIONAL CONSTRUCTION DETAILS SEE FACILITY NO. 040101, TM 5-301 THRU 5-303.

ERECTION PROCEDURES

1. LEVEL SITE.
2. PLACE WATERPROOFING MEMBRANE ON GROUND BELOW SHELTER.
2. PLACE ARCH SECTION.
3. PLACE REAR END WALL AND BRACE TEMPORARILY.
4. PLACE AND TEMPORARILY BRACE FRONT END WALL.
5. CONNECT, TIGHTEN, TIE CABLES FROM END WALL TO END WALL (SEE CABLE DETAIL SHEET 2).
6. PLACE AND CONNECT ENTRANCEWAY TO SHELTER (ENTRANCEWAY NOT INCLUDED IN THIS SET OF DRAWINGS)
3. PLACE WATERPROOFING MEMBRANE OVER SHELTER AND ENTRANCEWAY.
3. BACKFILL SHELTER TO DESIRED EARTH COVER BEING CAREFUL TO RAISE AND PACK EARTH FILL IN UNIFORM LIFTS OF EQUAL DEPTH ON OPPOSITE SIDES OF ARCH SECTIONS AND END WALL SECTIONS.

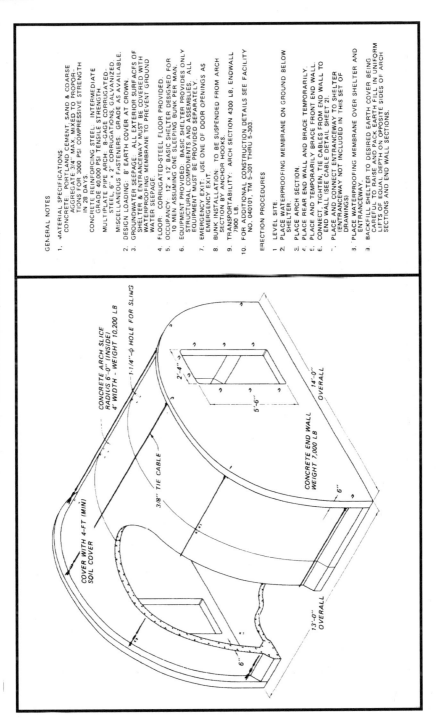

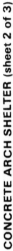
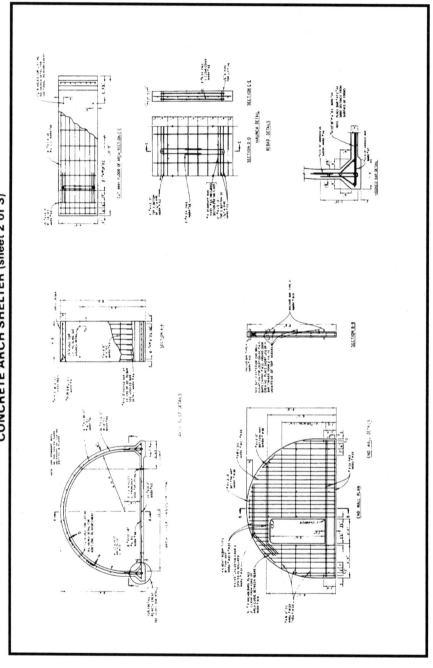

Position Design Details 147

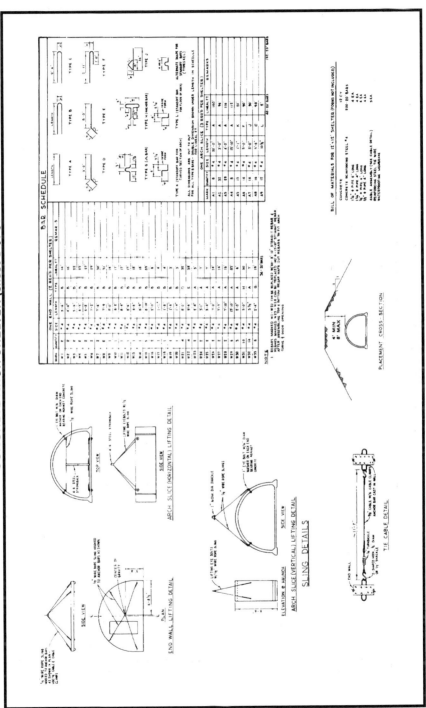

CONCRETE ARCH SHELTER (sheet 3 of 3)

METAL PIPE ARCH SHELTER (sheet 1 of 2)

GENERAL NOTES

1. MATERIAL SPECIFICATIONS:
   CONCRETE: PORTLAND CEMENT, SAND & COARSE AGGREGATE 3/4" MAX. MIXED TO PROPORTIONS FOR 3000 PSI COMPRESSIVE STRENGTH IN 28 DAYS.
   CONCRETE REINFORCING STEEL: INTERMEDIATE GRADE 40,000 PSI TENSILE STRENGTH.
   MULTIPLATE PIPE ARCH: 8-GAGE CORRUGATED STEEL, 6" x 2" CORRUGATIONS, GALVANIZED.
   MISCELLANEOUS FASTENERS: GRADE AS AVAILABLE.
2. DESIGN LOADING: 8' EARTH COVER AT CROWN.
3. GROUNDWATER SEEPAGE: ALL EXTERIOR SURFACES OF SHELTER & ENTRANCEWAY MUST BE COVERED WITH WATERPROOFING MEMBRANE TO PREVENT GROUND WATER SEEPAGE.
4. FLOOR: CORRUGATED-STEEL FLOOR PROVIDED.
5. OCCUPANCY: 12' x 12' BASIC SHELTER DESIGNED FOR 10 MEN ASSUMING ONE SLEEPING BUNK PER MAN.
6. EQUIPMENT PROVIDED: BASIC SHELTER PROVIDES ONLY STRUCTURAL COMPONENTS AND ASSEMBLIES. ALL EQUIPMENT MUST BE PROVIDED SEPARATELY.
7. EMERGENCY EXIT: USE ONE OF DOOR OPENINGS AS EMERGENCY EXIT.
8. BUNK INSTALLATION: TO BE SUSPENDED FROM ARCH SECTION BY ANCHOR HOOKS.
9. TRANSPORTABILITY: ARCH SECTION 4300 LB. ENDWALL 7900 LB.
10. FOR ADDITIONAL CONSTRUCTION DETAILS SEE FACILITY NO. 040201, TM 5-301 THRU 5-303.

ERECTION PROCEDURES

1. LEVEL SITE.
2. PLACE WATERPROOFING MEMBRANE ON GROUND BELOW SHELTER.
3. PLACE ARCH SECTION.
4. PLACE REAR END WALL AND BRACE TEMPORARILY.
5. PLACE AND TEMPORARILY BRACE FRONT END WALL.
6. CONNECT, TIGHTEN, TIE CABLES FROM END WALL TO END WALL (SEE CABLE DETAIL SHEET 2).
7. PLACE AND CONNECT ENTRANCEWAY TO SHELTER (ENTRANCEWAY NOT INCLUDED IN THIS SET OF DRAWINGS).
8. PLACE WATERPROOFING MEMBRANE OVER SHELTER AND ENTRANCEWAY.
9. BACKFILL SHELTER TO DESIRED EARTH COVER BEING CAREFUL TO RAISE AND PACK EARTH FILL IN UNIFORM LIFTS OF EQUAL DEPTH ON OPPOSITE SIDES OF ARCH SECTIONS AND END WALL SECTIONS.

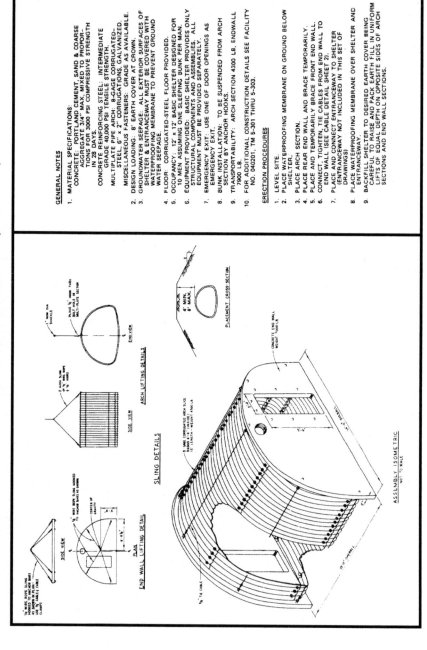

Position Design Details 149

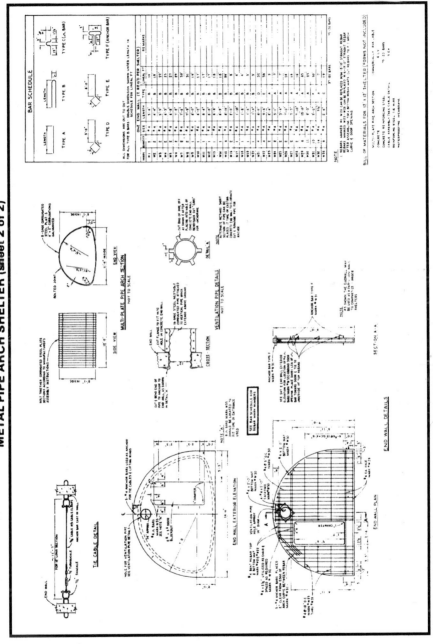

METAL PIPE ARCH SHELTER (sheet 2 of 2)

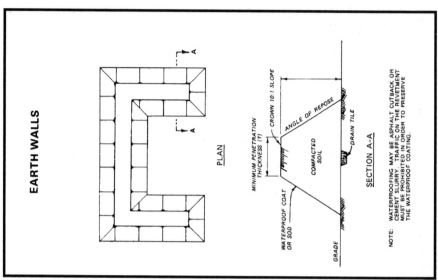
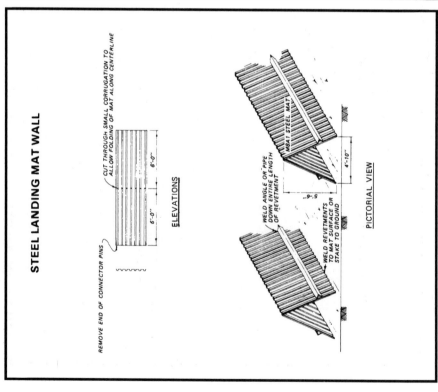

SOIL-CEMENT WALL

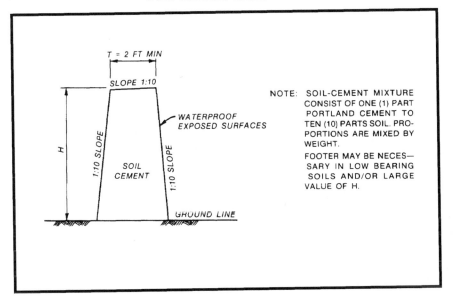

EARTH WALL WITH REVETMENT

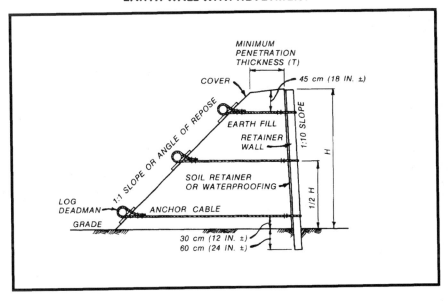

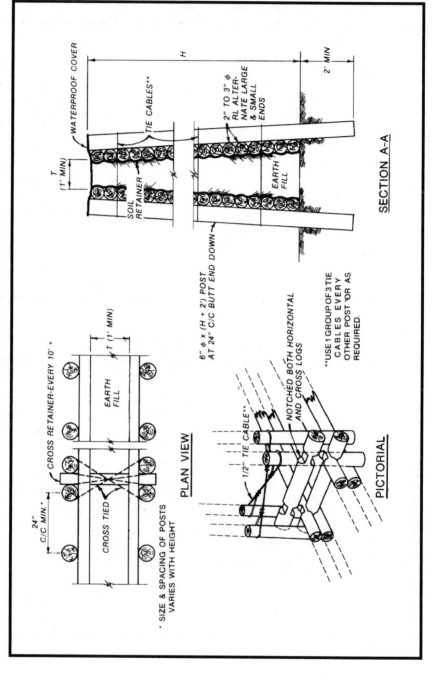

Position Design Details 153

SOIL BIN WALL WITH TIMBER REVETMENT

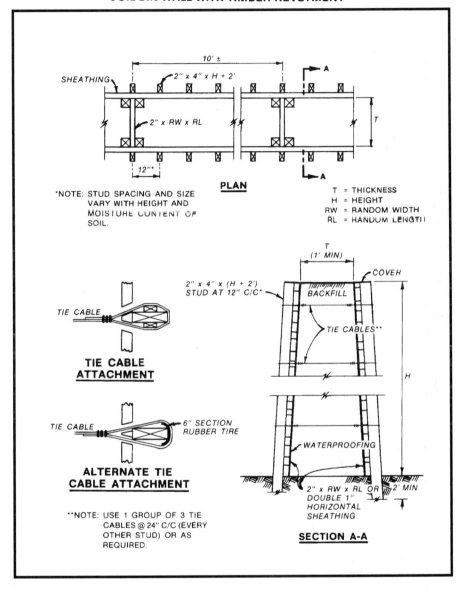

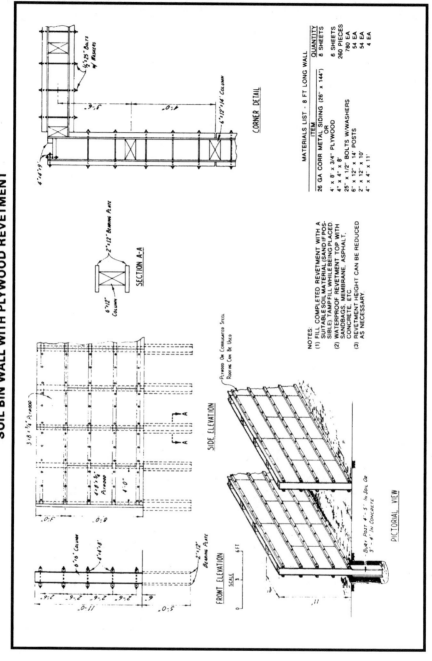

Position Design Details 155

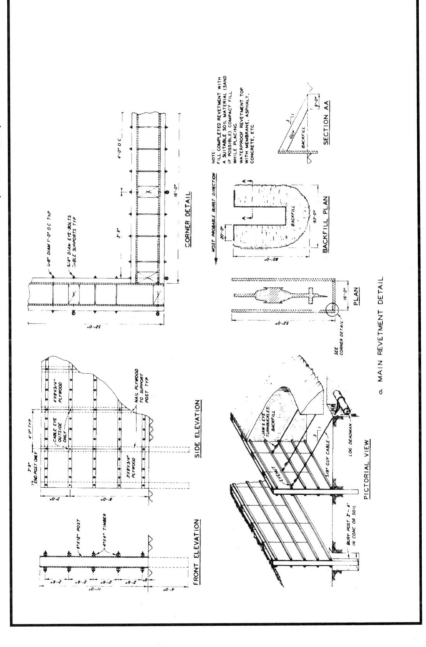

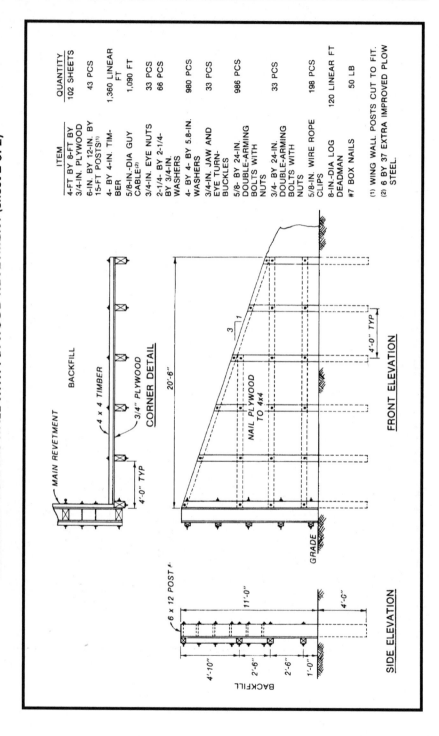

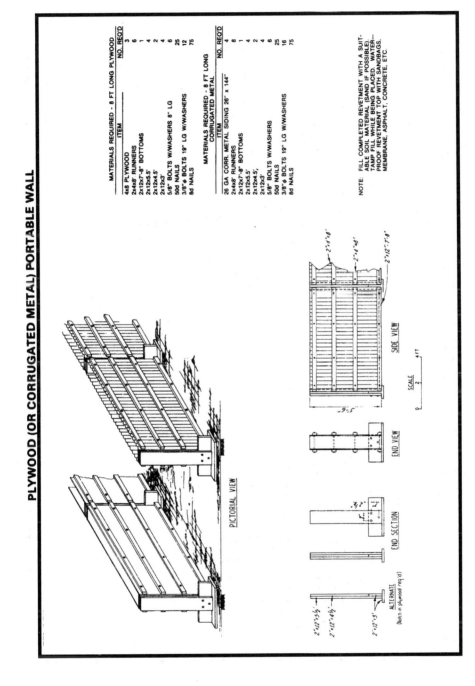

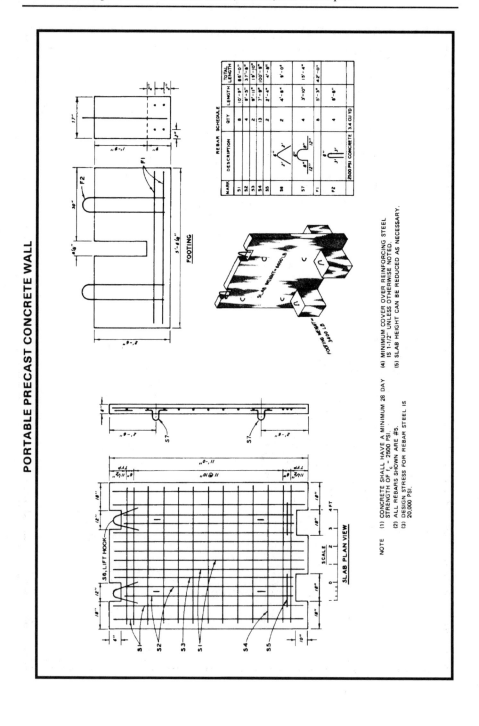

Position Design Details 159

CAST-IN-PLACE CONCRETE WALL

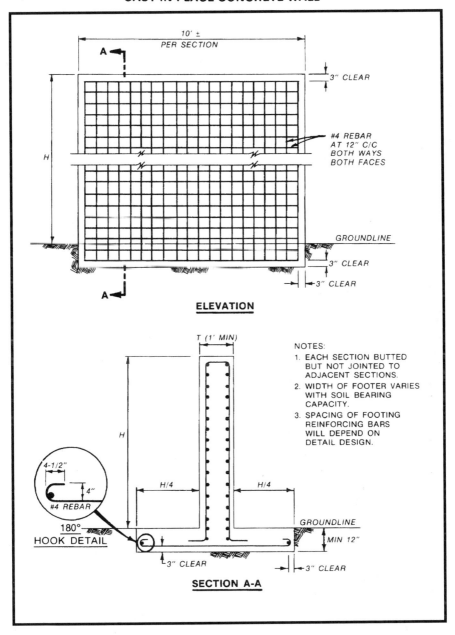

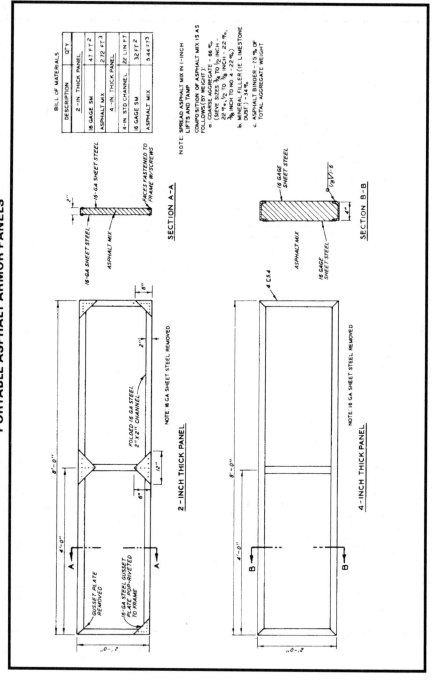

Position Design Details 161

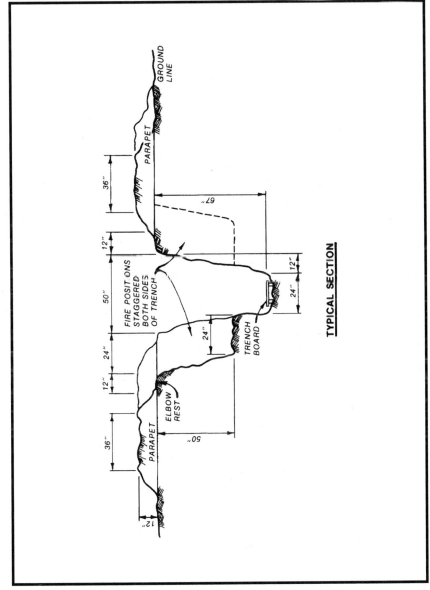

VEHICLE FIGHTING POSITIONS (DELIBERATE)

Vehicle Type	Position Dimension, ft [2] Length (A)	Width (B)	Hull Depth (C)[5]	Turret Depth (D)[5]	Weapon System Deflection	Weapon System Evaluation	Volume of Earth Moved (cy) Hull	Turret[6]	Total[7]	Equipment Hours [4] D7 Dozer/M9 ACE Hull	Turret[6]	Total[7]
DELIBERATE[1]												
M113 series carrier [3]	22	14	6	7½	-10°	—	69	124	193	0.6	1.0	1.6
M901 improved TOW vehicle	22	14	7	9	-10° gun	+30°	80	148	228	0.6	1.1	1.7
M2 and M3 fighting vehicle	26	16	7	10	-10° TOW	+60° +30°	108	218	326	0.8	1.7	2.5
M1 main battle tank	32	18	5½	9	-10°	+20°	118	268	386	0.9	2.0	2.9
M60 series main battle tank	30	18	6	10	-10°	+20°	120	278	398	0.9	2.1	3.0
M48 series battle tank	30	18	6	10	-10°	+20°	120	278	398	0.9	2.1	3.0

Notes:

1. Hasty positions for tanks, IFVs, and ITVs not recommended.
2. Position dimensions provide an approximate 3-foot clearance around vehicle for movement and maintenance and do not include access ramp(s).
3. Includes M132 flamethrower and M103 Vulcan.
4. Production rate of 100 bank cubic yards per .75 hour. Divide construction time by 0.85 for rocky or hard soil, night conditions, or closed hatch operations (M9). Ripper needed if ground is frozen. Use of natural terrain features will reduce construction time.
5. All depths are approximate and will need adjustment for surrounding terrain and fields of fire.
6. Turret volume (c) plus approach volume (b). Path length (E) is approximately ½(A).
7. Hull volume (a) plus approach volume (b) plus turret volume (c).

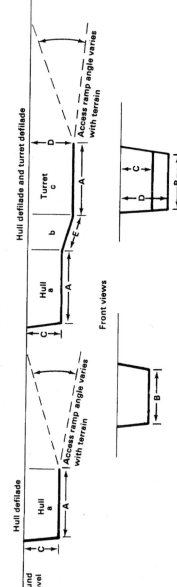